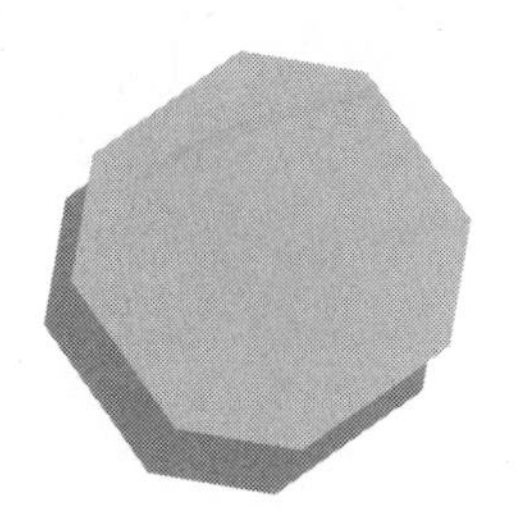

Geometry Experiments
Exploring Algebraic Connections

Mary Jean Winter
Ronald J. Carlson

Addison-Wesley Publishing Company
Menlo Park, California • Reading, Massachusetts • New York
Don Mills, Ontario • Wokingham, England • Amsterdam
Bonn • Milan • Madrid • Sydney • Singapore • Tokyo
Seoul • Taipei • Mexico City • San Juan

To the many teachers and students who have encouraged us, we'd like to say thank you. Our special thanks to Peggy Bosworth, Leigh Boyce, Peggy Cragin, Allan Everhart, Garth LaPointe, Lisa Ledwith, Jerry Neidenback, Dee Neidlinger, Ed Osuchowski, David Strohl, Bill Tabriski, and Laura Trausch for their constructive suggestions and comments. Thanks also to Alyson Elliott and Rebecca Froelicher. Very special thanks to our editor, Mali Apple.

Project Editor: Mali Apple

Production: Barbara Atmore

Design Manager: Jeff Kelly

Design: Detta Penna

Cover Art: Rachel Gage

ISBN 0-201-49345-4

2 3 4 5 6 7 8 9 10 - ML - 99 98 97 96 95

This book is printed on recycled paper.

Contents

Introduction

"A pump, not a filter." "Lean and lively." Although these phrases have been used to describe the spirit of the calculus-reform movement, they apply equally well to the intent of *Geometry Experiments.* In the past, calculus and geometry were often taught in a way that paid little attention to students' backgrounds, interests, and learning styles. A canonical body of abstract subject matter was presented to students who were expected to absorb the abstractions. Many calculus and geometry problems could be done by applying a template without necessarily understanding the mathematical concepts underlying the solution.

The spirit of reform in mathematics education is rising from the bottom—witness the curriculum changes at every level recommended by *The Curriculum and Evaluation Standards for School Mathematics* (copyright 1989 by the National Council of Teachers of Mathematics)—and receiving impetus at the top, as seen in changes in college algebra courses in which students use technology to define and make clear concepts, and in the many calculus projects that build mathematical knowledge from a concrete, real-world basis. At every level, the reforms reflect the agreement that there is room for all types of problem solvers and that students can and should use technological tools to construct their knowledge of mathematics.

Overview of the Geometry Experiments

The geometry experiments are based on concrete experiences: simple physical experiments in which students use familiar materials—boards, boxes, thread—to generate data, and calculators to plot data points quickly and accurately.

From the physical experiment, students make the links to mathematical concepts. On the Mathematical Analysis sheet for each experiment, students must

- represent the situation under investigation geometrically,
- identify algebraic relationships, and
- determine an underlying function.

By graphing the resulting function on calculators, students verify or reject their solutions. The use of calculators does more than replace graphing by hand; it helps students confirm the correctness of their deductions.

The Collect the Data sheet shows the Key Concepts—the statements of the algebraic relationship between the variables and related diagrams. The needs of classes vary; if you wish, you can conceal the Key Concepts before reproducing the Collect the Data sheet.

In the third part of each experiment, Interpret Your Findings, students use technology, algebra, and physical objects to extend their understanding of the mathematics concepts embedded in the experiment.

The Mathematical Analysis and Interpret Your Findings sheets provide opportunities for students who use many learning styles—from those who proceed by logical steps to those who need to verify what they know intuitively. By completing the experiments, students gain confidence and skills to investigate their own ideas—in other words, to become active learners.

Design of the Experiments

Each experiment is divided into three interrelated parts: Collect the Data, Mathematical Analysis, and Interpret Your Findings. There is a clear progression from the concrete level to algebraic/geometric representation, and to even more abstract and conceptual considerations:

I Collect the Data	Do the physical experiment
	Plot points
II Mathematical Analysis	Describe geometrically
	Represent algebraically
	Find general function
	Verify for specific experiment
III Interpret Your Findings	Investigate properties of the function
	Relate properties to the geometry of the original experiment
	Vary the experiment—make it dynamic

This progression is not one-directional. Throughout the experiment students go back and forth between geometry and algebra, between concrete objects and symbolic representations, between physical experience and theoretical speculations. This movement is made possible through the use of the graphing calculator.

Collect the Data—Performing the Experiment

Each experiment begins with a hands-on experience using easily obtained, familiar objects as equipment. To carry out the experiments, students must understand the basic geometric concepts of length, area, and angle, and how they are measured or calculated. The variables must be identified and measured. In many experiments, y is not directly measured, but must be calculated. Once the data points are determined, students plot them on a graphing calculator.

Key Concepts for the experiment are listed on the Collect the Data page. The Key Concepts include equations relating the variables and the constants of the experiment and relevant diagrams. If the Key Concepts are not appropriate for the level of your class, conceal them with a self-stick note before copying the Collect the Data page.

Mathematical Analysis

As the outline on page 2 indicates, the Mathematical Analysis part of an experiment has many components. First, students must label the geometric components of the experiment. Quantities that remain fixed throughout the experiment are represented by constants; for example, if height is held constant, in the analysis it is referred to as h. For some students, working with several symbols at a time will be challenging.

Students then use their knowledge of geometrical relationships to represent the experiment algebraically. The principal relationships needed are those from similar triangles and the Pythagorean Theorem. Then students use algebraic skills to solve for the dependent variable, y, as a function of x (and of all the constant quantities).

Once the general function—that is, an equation for y—has been found, students replace all the constants with their actual values. On the same calculator that shows the data points, students draw the graph of "their" function. Watching the graph "hit" the data points is an important part of the learning experience. Visual confirmation provides both immediate feedback and reinforcement.

If the function is clearly not representative of the points, students need to check both their understanding of the physical experiment (they may have measured, or calculated, the wrong quantity) and their mathematical analysis. Sometimes students will need to convert the units they have recorded to make the graph "behave."

Interpret Your Findings—Making Connections

At this stage students are no longer working with data points, but with the function whose validity they have confirmed. This part contains a set of questions designed to strengthen and deepen students' understandings of their knowledge. For example, which value of x leads to the maximum area? What does that mean in terms of the original experiment? Students frequently return to the experimental situation to examine one instance more deeply.

Other questions relate the experimental situation to the function: What are the physical meanings of the constants? If they change, how does the function change? How does the graph change? For such questions, students may find it easier to use the unsimplified form of the expression.

Students occasionally ask questions about the "path" followed by one particular point. Sometimes the coordinates of that point depend on the independent variable. In that case, students can examine the path geometrically by graphing the coordinates in the parametric mode. Sometimes students must use geometric definitions to help them develop algebraic representations that describe the conditions of the experiment, such as: "If the distance is a constant, there must be a circle involved." They use their algebraic skills to describe the locus of points.

Each experiment gives rise to different types of questions. Some questions are intended to be answered through the use of a graphical method, some through algebraic analysis and manipulations, and some by the use of tabular results. Some questions can be answered by using any of several methods. A student's approach will depend on his or her experiences, knowledge, and personal preference. As they proceed, students will find that algebra makes geometry dynamic. From a theoretical viewpoint, varying the experiment can be regarded as introducing an algebraic variable.

Teaching Notes

Each experiment is preceded by a set of Teaching Notes, which contains the following information:

- Nature of the experiment, including the independent and dependent variables
- Equipment list
- Procedure to follow
- Mathematical Analysis: a detailed description of the function to be found

The Teaching Notes may also contain the following:

- Suggestions for extensions or follow-up questions
- Procedure for organizing and analyzing class results
- Algebraic or experimental results to be found in the experiment

Cooperative Learning and Individual Accountability

The experiments combine the benefits of cooperative learning with individual accountability and the use of technology. Each experiment involves student interaction and discussion, informal peer instruction, shared ownership of a completed project, the pride of a completed project, and the satisfaction of personally understanding the mathematics.

Most of the experiments require two or more people to successfully accomplish the setup and the collection of data. The use of a cooperative group also helps to ensure the quality of the data. After collecting the data together, the group has a real interest in verifying that their function fits their data. There will be discussion and informal peer teaching: "I think the maximum point will be. . . ." "The x-intercept should represent the. . . ." The extension questions also produce discussions and conjectures to be challenged and defended: "I think the maximum value represents. . . ." "Why did we get a parabola for our function?"

You may want to have cooperative groups present their project to the entire class. While this type of activity is new to many math classrooms, it provides students an opportunity to strengthen their oral communication skills.

Each student should write up every experiment completely, describing in writing and with diagrams the exact nature of the experiment, and carrying out all of the steps involved in the Mathematical Analysis and Interpret Your Findings sheets. Every student will then have a complete record of the experiments as demonstration of accomplishment and for possible inclusion in a portfolio.

Required Concepts and Skills

The experiments can be used in any geometry course (whether preceded by one or by two years of algebra), in an integrated mathematics curriculum that combines algebra and geometry, or in a precalculus course. Students who are successful with the standard presentation of mathematics—manipulation and memorization—find the experiments both a welcome change and an unexpected link between geometry and algebra.

To complete an experiment successfully, a student should already have studied the underlying geometric concepts and possess the algebra skills necessary to obtain the function. The table on the next page provides a quick overview of the required student background in geometry and algebra.

Required Concepts and Skills

Experiment	*Geometry Concepts and Formulas*	*Algebra Skills*
1 *Perimeter of a Rectangle*	area and perimeter of rectangle	substitution solving equations
2 *Area of an Equilateral Triangle*	area of triangle Pythagorean Theorem	substitution solving equations
3 *Area of a Rectangle*	area and perimeter of rectangle	substitution solving equations
4 *Area of an Isosceles Triangle*	area of triangle Pythagorean Theorem	substitution solving equations
5 *Ramp Height*	Pythagorean Theorem	substitution solving equations
6 *Pet Leash*	Pythagorean Theorem	substitution solving equations
7 *Casting Shadows*	similar triangles Pythagorean Theorem	solving rational equations
8 *Stretching Points*	similar triangles	solving rational equations
9 *Little House*	area and perimeter of rectangle area of triangle Pythagorean Theorem	(algebra is easier if Experiments 3 and 4 have been done)
10 *Reflections*	measuring angles sum of exterior angles of polygons	substitution solving equations
11 *Looking Down*	similar triangles	solving rational equations
12 *Polygons*	supplementary angles	solving equations
13 *Stars*	supplementary angles	solving equations
14 *Stretchy Isosceles Triangle*	similar triangles Pythagorean Theorem	solving rational equations
15 *School Flower Beds*	area of rectangle Pythagorean Theorem	substitution solving equations
16 *Scaling the Wall*	Pythagorean Theorem parametric equations similar triangles	plotting x as a function of y (parametric equations)
17 *Class Photo*	law of cosines midpoint of a line segment	solving equations
18 *Sliding Down*	distance formula Pythagorean Theorem	substitution solving equations
19 *Squashed Boxes I*	area of parallelogram sin x	right triangle trigonometric functions
20 *Squashed Boxes II*	law of cosines properties of parallelograms	solving equations

The NCTM Standards for Geometry

The Curriculum and Evaluation Standards for School Mathematics (copyright 1989 by the National Council of Teachers of Mathematics) provides a vision of mathematical literacy for a changing world and establishes guidelines to help revise mathematics curriculum to reflect the new definition of mathematical literacy. *Geometry Experiments* reflects the same spirit and addresses the same aims by implementing the goals espoused in the *Standards:* lifelong learning, mathematically literate workers, and opportunity for all students. Learning to value mathematics, becoming confident mathematical problem solvers, and learning to communicate and reason mathematically are outcomes that can be attained through the use of the experiments.

New Standards for Mathematical Content

Two of the new geometry standards for students in grades 9 to 12 are especially relevant:

- *Standard 7:* Geometry from a Synthetic Perspective. "The interplay between geometry and algebra strengthens students' ability to formulate and analyze problems from situations both within and outside mathematics."
- *Standard 8:* Geometry from an Algebraic Perspective. "Physical models and other real-world objects should be used to provide a strong base for the development of students' geometric intuition so that they can draw on these experiences in their work with abstract ideas."

These standards include the following goals for students in grades 9 to 12:

- Represent problem situations with geometric models and apply properties of figures
- Classify figures in terms of congruence and similarity and apply these relationships
- Deduce properties of, and relationships between, figures from given assumptions
- Translate between synthetic and coordinate representations
- Deduce properties of figures using transformations and using coordinates

Changes in Instructional Practices

In addition to changes in the curriculum, the *Standards* include the following goals for changes in instructional practice:

- The active involvement of students in constructing and applying mathematical ideas
- Problem solving as a means as well as a goal of instruction
- The use of a variety of instructional formats (small groups, individual ex-

plorations, peer instruction, whole class, discussion, project work)

- The use of calculators and computers as tools for learning and doing mathematics
- Student communication of mathematical ideas, orally and in writing

Evaluation and Assessment

The experiments can be used in a variety of ways: for enrichment, for motivation, for reinforcement, or to introduce a topic. They can be used with students of varied backgrounds and abilities. The method chosen for evaluation and assessment will depend on how the experiment is used. An experiment can be graded and returned to the student, or it might be placed in the student's portfolio. Some suggestions for evaluation, assessment, and examples of assessment questions follow.

Suggestions for Evaluating the Experiments

Each experiment has three parts—Collect the Data, Mathematical Analysis, and Interpret Your Findings. The data collection and analysis parts are almost self-checking. If the graph does not pass near the data points, students can see that something is wrong and make the necessary adjustments to the function or equation. This process allows students to verify their work and to make connections between the experiment, the function, and the data. A student might need more than one attempt to find the correct function.

A suggested weighting for the first and second halves of the experiment is 40 percent for the Collect the Data and Mathematical Analysis and 60 percent for the Interpret Your Findings questions. Since the function must be found before proceeding to the questions, this means every student will build on a basis of 40 points. Depending on their background, some Interpret Your Findings questions might not be appropriate for your students. Feel free to delete or replace such questions.

Alternatively, you can create a summary page, indicating for each part whether the student's work is exemplary, satisfactory, or unsatisfactory. Include comments on the student's level of understanding.

Assessment Alternatives

The experiments provide a vehicle for students to demonstrate their understanding of geometric concepts and their connections to other areas of mathematics such as trigonometry and algebra. Besides completing the *product*—the experiment worksheets—students are given a chance to model the *process*—the type of inquiry and investigation that mathematicians find valuable. It is important that both the product and the process are evaluated and assessed. Because of the hands-on activities, teachers should anticipate and expect more student involvement as the students themselves begin to pose questions such as, "What happens if we double the length of the side?" These inquiries provide opportunities for other means of assessment.

Assessing the Product: Reports and Portfolios

The experiments are well suited for a class report or for inclusion in a student's portfolio. Completion of the Collect the Data sheet shows that the student understands the experiment and can communicate that understanding in writing. Completion of the Mathematical Analysis sheet demonstrates that the student understands the relevant geometry and possesses the necessary algebraic skills, and it also provides evidence of higher-order thinking skills.

You might ask students who have done several experiments to select one or two write-ups for inclusion in their portfolios. They should write an introduction to the selected experiment(s) to explain their methods, the mathematical concepts they used, and the significance of their findings.

Other means of assessing include having a cooperative team present an experiment as if to a meeting of mathematicians. The team should explain the experiment, describe their data and analysis, and discuss their answers to some of the Implications questions. At the end of the presentation, the team should pose one or two additional questions or extensions for "further research."

If the software and hardware is available, a cooperative team of students might demonstrate the experiment using programs such as *Geometer's Sketchpad* or *Cabri.* The demonstration should be followed by an explanation of how the experiment was implemented.

Assessing the Process: Observation

As students work on the experiments, you can observe each group with these questions in mind:

- Are the partners focused on the experiment?
- Is each student contributing?
- What is the nature of the conversations—are students exchanging facts and discussing results?
- What level of understanding do the discussions show?
- Are students observing and analyzing trends?

Note any interesting or informative student comments.

Assessing the Process: Questions

Four basic types of questions can be used to test students' understanding; examples are given here. The Teaching Notes for many of the experiments suggest other questions. Many extension questions can be adapted for computer investigation.

Experiment-based Questions

1. Malcolm did Experiment 19, Squashed Boxes I, but after obtaining the graph and deriving the equation, Area = 12 sin x, he lost his box without recording the base and the height. Can he tell from the equation the exact size of his box? If not, what are three different boxes he might have used?
2. *(Ask the student to re-create an experiment mentally.)* In Experiment 3, Area of a Rectangle, Delia has several rectangles of area 124. Find the perimeter of such a rectangle as a function of the length of the base.
3. With the same mirror, Keisha and Christine both collected data for Experiment 11, Looking Down. Each did her own analysis and found the equation of a line. Keisha is noticeably taller than Christine. How do their lines compare? Does it matter where the mirror is placed?

Graph-based Questions

1. In Experiment 1, Perimeter of a Rectangle, Chip and Sandy both found the perimeter of a rectangle of fixed area. Sandy's rectangles had an area of 10, while Chip's had an area of 20. Here is the graph of Sandy's perimeter function. Sketch the graph of Chip's perimeter function.

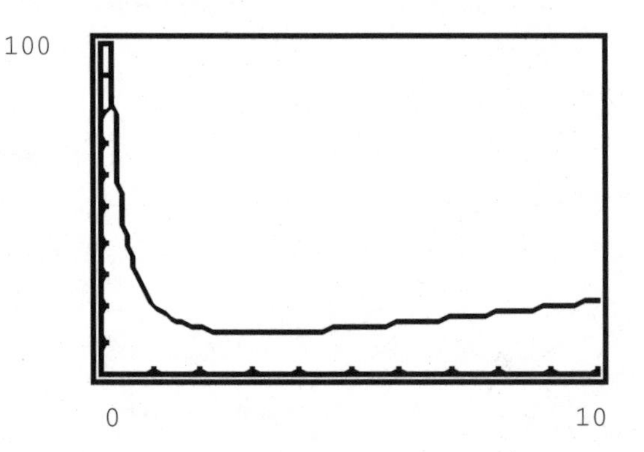

2. You did Experiment 3, Area of the Rectangle, and obtained this graph. Using a different perimeter for her rectangles, Zia did the same experiment. Can you determine the perimeter of Zia's rectangles? Explain your method.

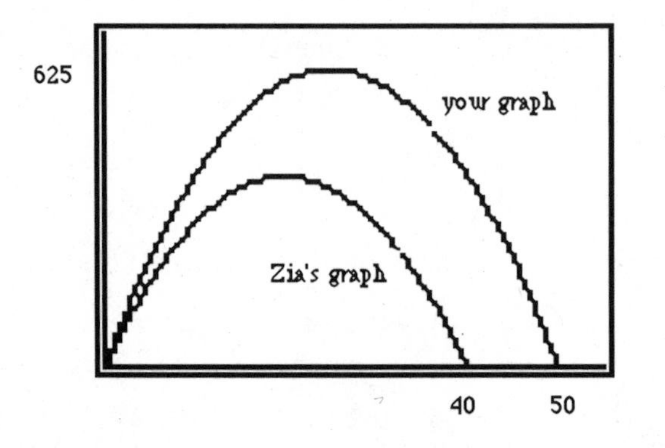

Equation-based Questions

1. Jordan found this equation, $y = \sqrt{x^2 + 900}$, for Experiment 6, Pet Leash. He recorded his measurements in inches. Find the equation for the same experiment if Jordan had recorded the measurements in centimeters.
2. *(Ask an abstract question that has already been considered concretely.)* In Experiment 2, Area of an Equilateral Triangle, suppose an equilateral triangle has a side x. Find the area of the triangle as a function of x.

Open-ended (What-if) Questions

1. After doing Experiment 20, Squashed Boxes II, Alex wants to find the length of the other diagonal as a function of x. Explain how Alex can derive the new equation without having to redo the experiment .
2. Suppose a different class did Experiment 2, Area of an Equilateral Triangle, but they used centimeters and your class used inches. Can both data sets be combined? Why or why not? Explain your answer.

Below is sample student work for Experiment 4, Area of an Isosceles Triangle.

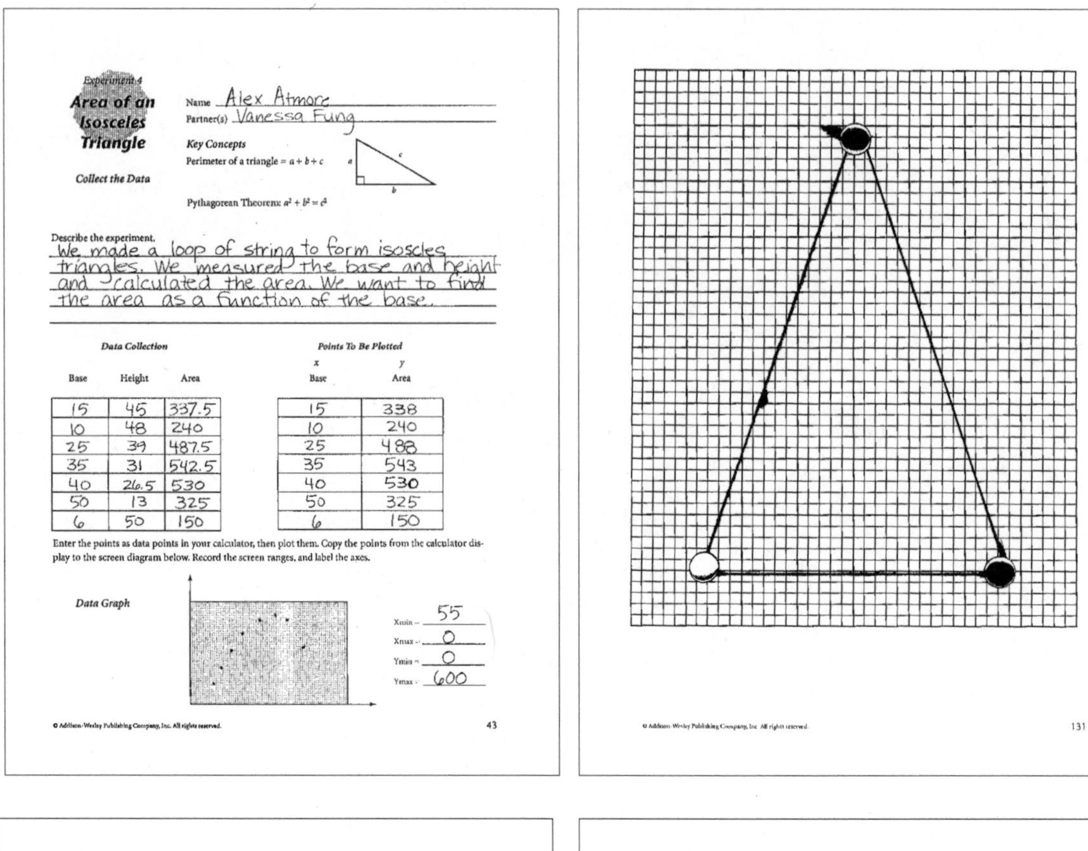

Experiment 4
Area of an Isosceles Triangle

Collect the Data

Name Alex Atmore
Partner(s) Vanessa Fung

Key Concepts
Perimeter of a triangle = $a + b + c$

Pythagorean Theorem: $a^2 + b^2 = c^2$

Describe the experiment.
We made a loop of string to form isoscles triangles. We measured the base and height and calculated the area. We want to find the area as a function of the base.

Data Collection

Base	Height	Area
15	45	337.5
10	48	240
25	39	487.5
35	31	542.5
40	26.5	530
50	13	325
6	50	150

Points To Be Plotted

x Base	y Area
15	338
10	240
25	488
35	543
40	530
50	325
6	150

Enter the points as data points in your calculator, then plot them. Copy the points from the calculator display to the screen diagram below. Record the screen ranges, and label the axes.

Data Graph

Xmin = 55
Xmax = 0
Ymin = 0
Ymax = 600

© Addison-Wesley Publishing Company, Inc. All rights reserved. 43

© Addison-Wesley Publishing Company, Inc. All rights reserved. 131

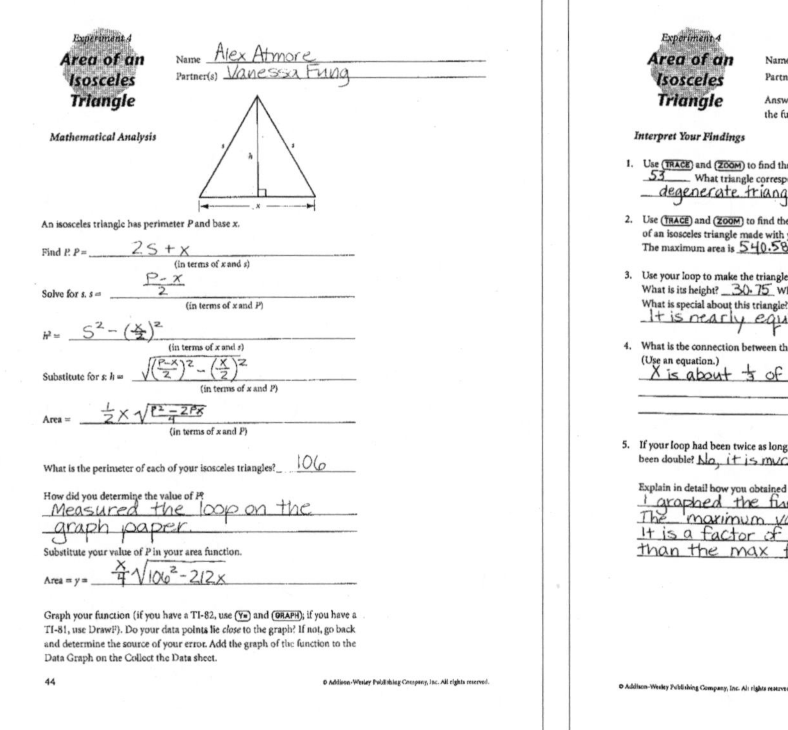

Experiment 4
Area of an Isosceles Triangle

Mathematical Analysis

Name Alex Atmore
Partner(s) Vanessa Fung

An isosceles triangle has perimeter P and base x.

Find P. $P =$ $2s + x$ (in terms of x and s)

Solve for s. $s =$ $\frac{P-x}{2}$ (in terms of x and P)

$h^2 =$ $s^2 - (\frac{x}{2})^2$ (in terms of x and s)

Substitute for s. $h =$ $\sqrt{(\frac{P-x}{2})^2 - (\frac{x}{2})^2}$ (in terms of x and P)

Area = $\frac{1}{2}x\sqrt{\frac{P^2-2Px}{4}}$ (in terms of x and P)

What is the perimeter of each of your isosceles triangles? 106

How did you determine the value of P?
Measured the loop on the graph paper.

Substitute your value of P in your area function.

Area = y = $\frac{x}{4}\sqrt{106^2 - 212x}$

Graph your function (if you have a TI-82, use Y= and GRAPH; if you have a TI-81, use Draw F). Do your data points lie *close* to the graph? If not, go back and determine the source of your error. Add the graph of the function to the Data Graph on the Collect the Data sheet.

44 © Addison-Wesley Publishing Company, Inc. All rights reserved.

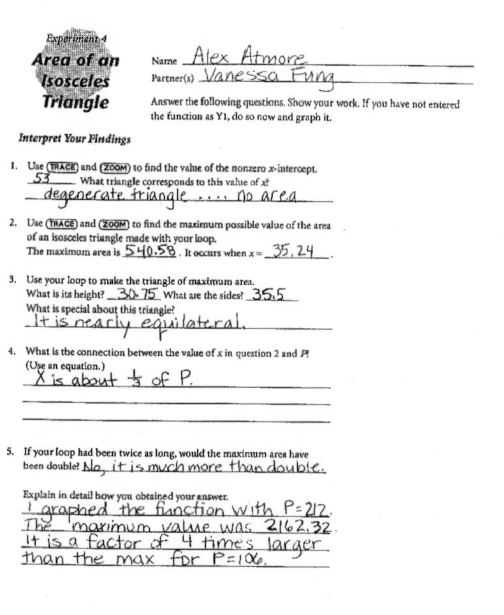

Experiment 4
Area of an Isosceles Triangle

Name Alex Atmore
Partner(s) Vanessa Fung

Answer the following questions. Show your work. If you have not entered the function as Y1, do so now and graph it.

Interpret Your Findings

1. Use TRACE and ZOOM to find the value of the nonzero x-intercept. 53 What triangle corresponds to this value of x?
   degenerate triangle .... no area
2. Use TRACE and ZOOM to find the maximum possible value of the area of an isosceles triangle made with your loop.
   The maximum area is 540.58. It occurs when $x =$ 35.24.
3. Use your loop to make the triangle of maximum area.
   What is its height? 30.75 What are the sides? 35.5
   What is special about this triangle?
   It is nearly equilateral.
4. What is the connection between the value of x in question 2 and P? (Use an equation.)
   X is about $\frac{1}{3}$ of P.
5. If your loop had been twice as long, would the maximum area have been double? No, it is much more than double.

   Explain in detail how you obtained your answer.
   I graphed the function with P=212. The maximum value was 2162.32. It is a factor of 4 times larger than the max for P=106.

© Addison-Wesley Publishing Company, Inc. All rights reserved. 45

Graphing Calculators and the Experiments

Technology—specifically, the graphing calculator—is the link between the experimental data students collect and the Mathematical Analysis. If the predicted curve agrees with the observed data, the experimental procedure and analysis have been validated. This validation, which requires using the calculator simply as a tool to plot points and draw the graph of a function, must be made before students proceed to the implications of their analyses.

Once the underlying function has been determined, additional features of the calculator are used to investigate the Implications of the result. The trace operation (which moves the cursor from one point to the next along a function) is mandatory; the zoom operation (which adjusts the viewing window) is convenient. The experiments assume that students have calculators with the capabilities outlined below.

Required Capabilities of Calculators

Procedure	***Required Technology***
Collect the Data	Plot points Replotting should be possible
Mathematical Analysis	Draw function without erasing points Plot in parametric mode (needed in Experiment 16 only)
Interpret Your Findings	Graph two or three functions Trace, find intersections, modify functions

It is also assumed that students know how to set appropriate screen ranges, graph a function, and use the trace and zoom operations. The experiments can be performed using any graphing calculator with these capabilities. Specific instructions are given here for the TI-81 and TI-82. The experiment pages—Teaching Notes, Collect the Data, Mathematical Analysis, and Interpret Your Findings—contain directions specific to these Texas Instruments calculators. If students are using other brands, appropriate changes will need to be made.

Plotting Points as Statistical Data: Using Graphing Calculators

Some students may be unfamiliar with using the statistical mode. Below are specific instructions for the TI-81 and TI-82. (▼) and (▶) are cursor controls; (Y=) is found in the upper left-hand corner.

The Texas Instrument TI-82 Graphing Calculator

Before beginning, clear the screen, the statistical memory lists L1 and L2, the *y* variables, and the statistical plots. One way to accomplish this is by pressing:

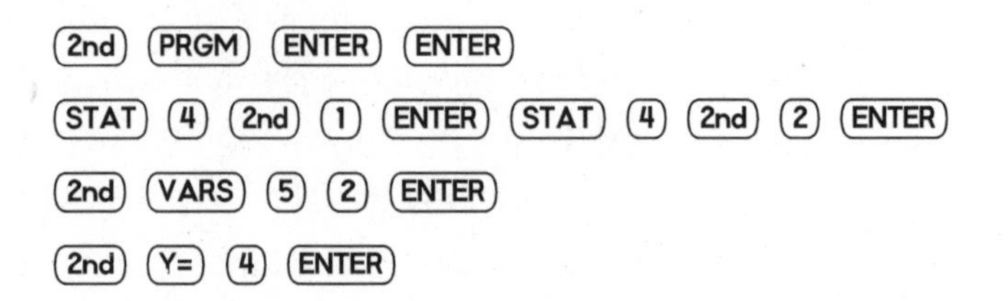

Check that the modes are set as shown:

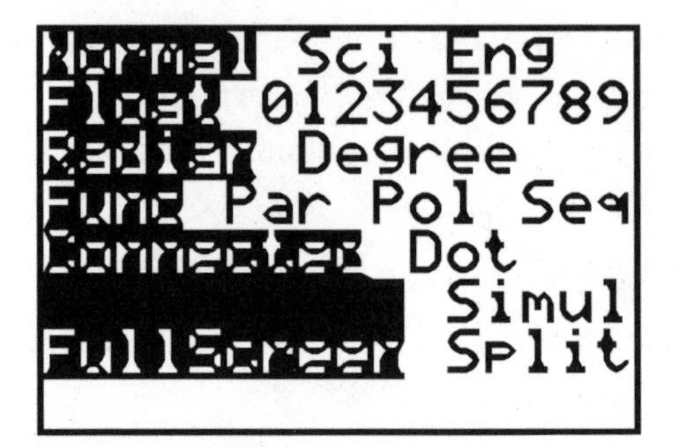

Set appropriate screen ranges (the first quadrant is standard for most experiments):

WINDOW FORMAT
Xmin=0
Xmax=5
Xscl=1
Ymin=0
Ymax=60
Yscl=10

Enter the data. The *x*-coordinates are entered as List 1; the *y*-coordinates are entered as List 2. To enter the points (1, 10), (2, 24), (3, 34), (4, 55), press:

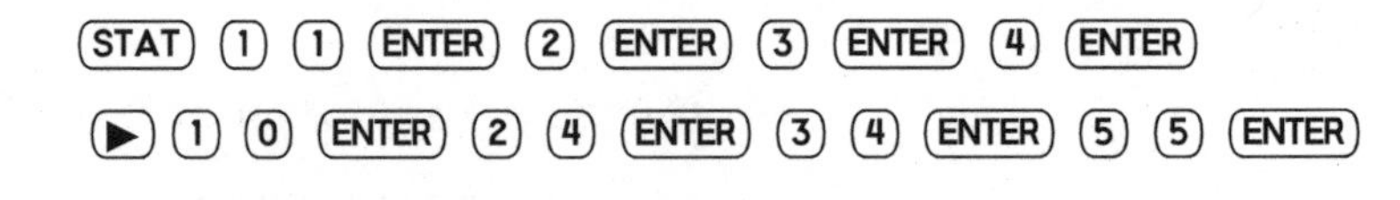

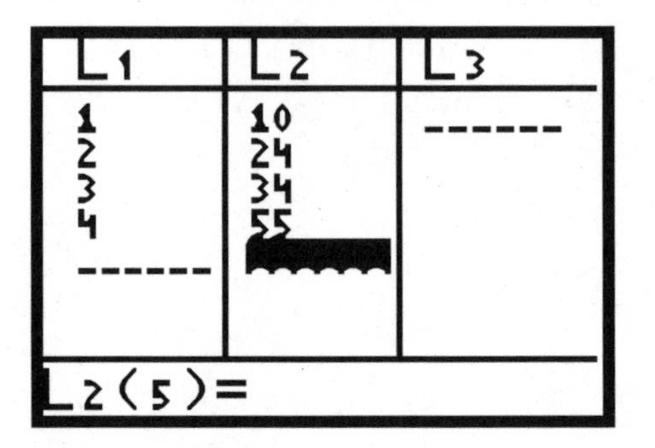

To draw a scatter plot, first press:

2nd Y= ENTER

This produces the screen describing the statistical plots and selects Plot1.

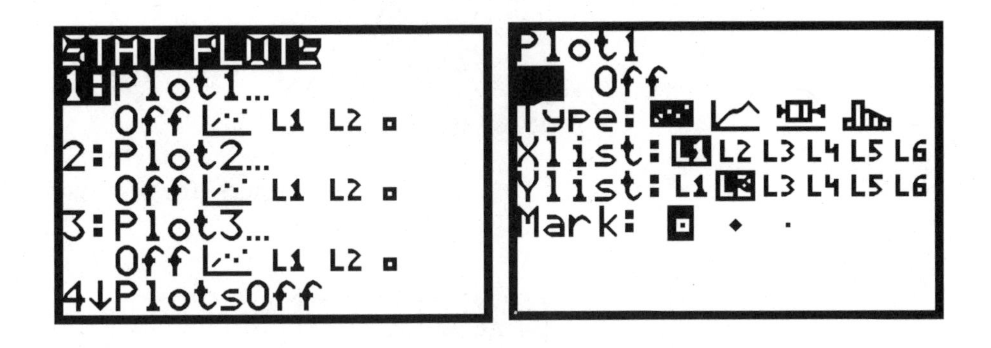

Select each of the following options by highlighting it and pressing ENTER

On

Type: Scatter

Xlist: L1

Ylist: L2

Mark □

To plot the points, press GRAPH:

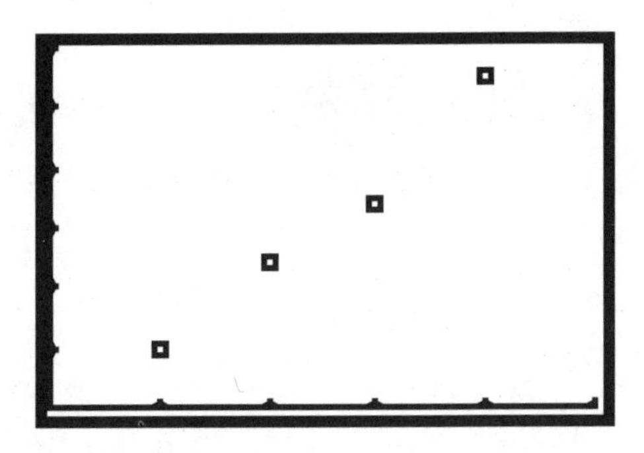

Graph a function on top of the points by entering the function in Y1 and pressing GRAPH.

The Texas Instruments TI-81 Graphing Calculator

Before beginning, clear the screen, the functions on the graph menu, and the statistical memory by pressing:

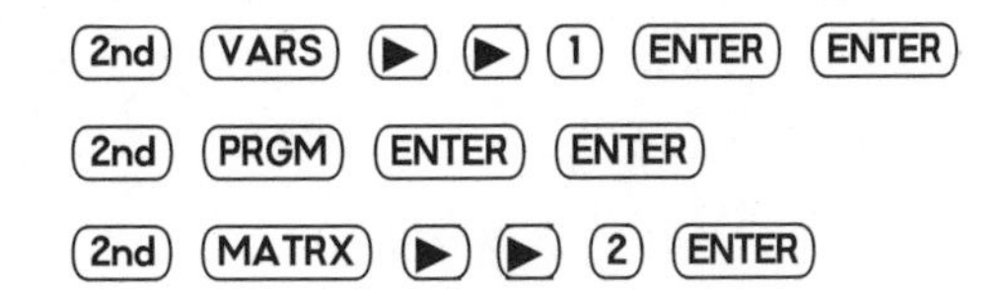

Check that the modes are set as shown:

Set appropriate screen ranges (the first quadrant is standard for most experiments):

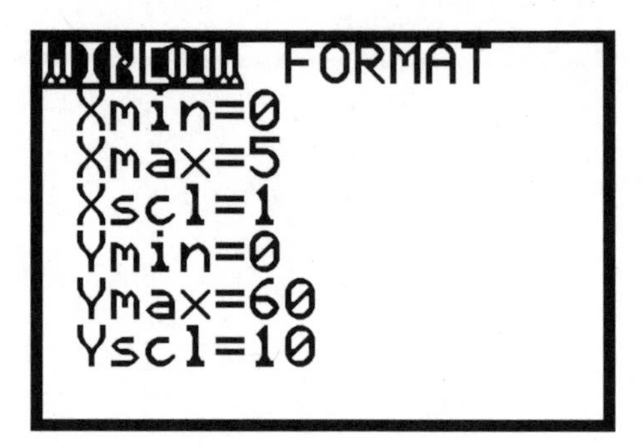

Plot the data points (which can be replotted if necessary). This is usually easiest when the coordinates of the points are entered as data in the statistical mode. To enter statistical mode and plot the first point—(1, 10) for example:

(2nd) (MATRX) (▶) (▶) (1) (1) (ENTER) (1) (0) (ENTER)

When the last point—(4, 55) for example—has been entered, draw a scatter plot of the points by pressing:

(2nd) (MATRX) (▶) (2) (ENTER)

To draw a graph of a function—$y = 3x$ for example—without erasing the points on the screen, use DrawF on the Draw menu by pressing:

(2nd) (PRGM) (6) (3) (X|T) (ENTER)

If the screen becomes too cluttered or if you make a mistake, clear the screen and replot the points by pressing:

(2nd) (PRGM) (1) (ENTER)

and redraw the scatter graph.

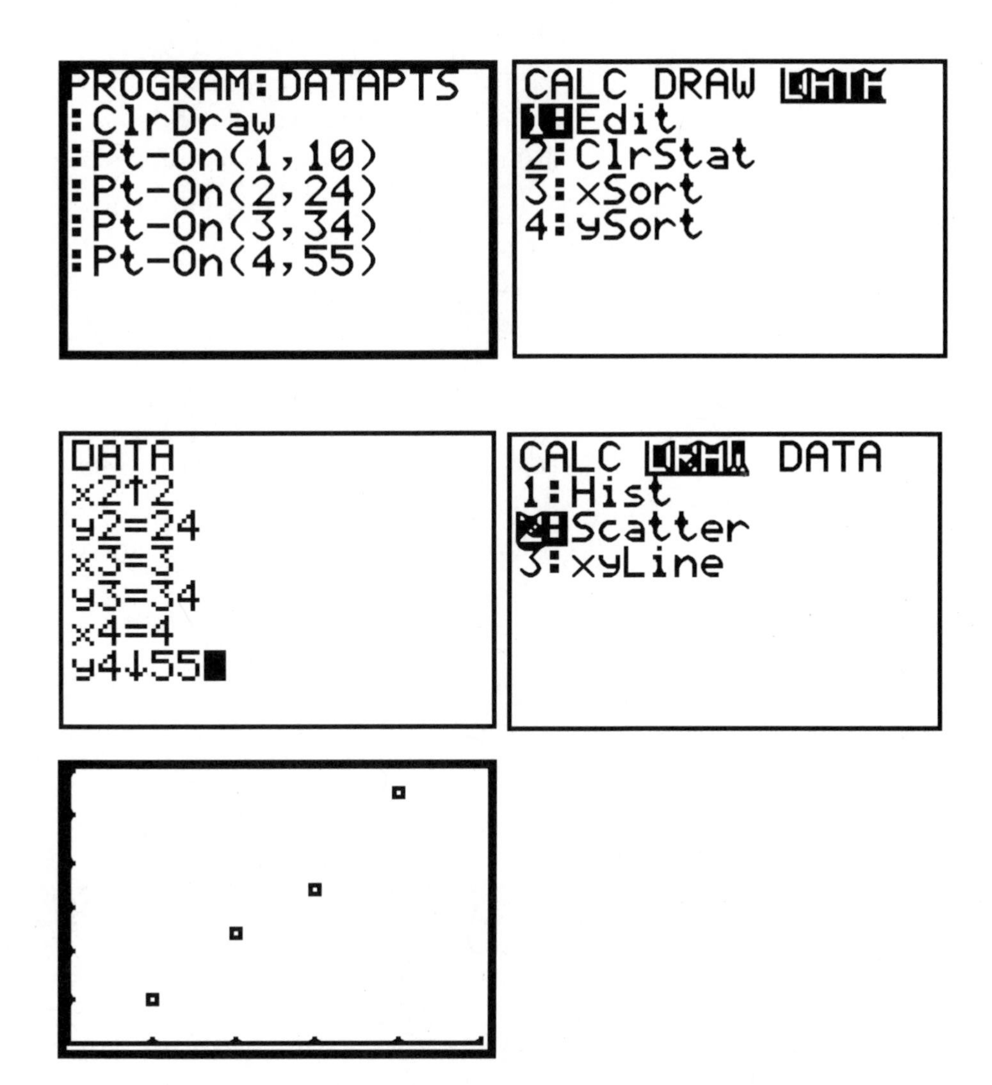

Introduction to Using Parametric Graphs (How to Graph $x = y^2$)

Parametric equations were formerly usually introduced at the end of Algebra II, when students first studied the trigonometric functions sin t and cos t. Graphing the equations $x = \cos t$, $y = \sin t$ produces a graph of a circle that looks continuous, while the graphs of $y = \sqrt{1 - x^2}$ and $y = -\sqrt{1 - x^2}$ do not produce a continuous graph. Gaps in the circle appear near the points $x = 1$ and $x = -1$.

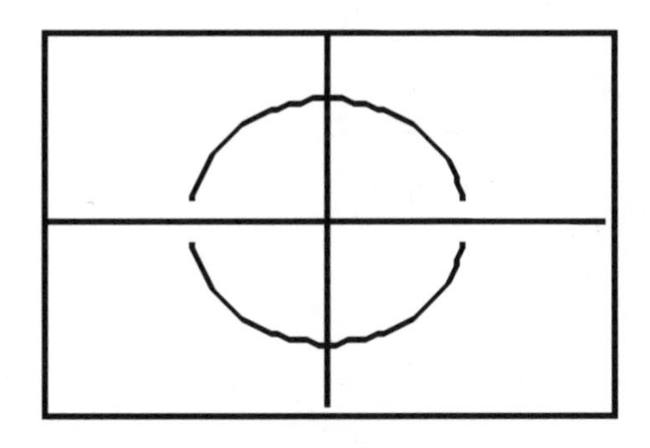

Graphing calculators make it easier to introduce parametric equations earlier in the curriculum.

A Simple Example

Suppose a person takes a walk on the Cartesian plane, starting at (0, 0), and that the person's position is recorded at every second:

t (seconds)	*x*-coordinate	*y*-coordinate
0	0	0
1	4	2
2	8	4
3	12	6

Note that at every second, the person goes twice as far in the *x* direction as in the *y* direction. What does the path look like? We could just plot the points (x, y), but if the motion were more complicated, that wouldn't be sufficient.

The person's coordinates (x, y) as functions of t are $x = 4t$ and $y = 2t$. To see the graph of the path, select parametric mode and graph the functions:

$X_{1T} = 4T$

$Y_{1T} = 2T$

Set the range as shown, and then graph

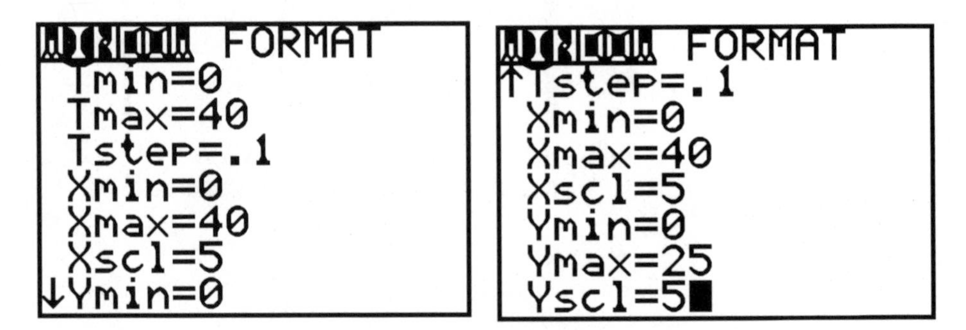

Range for the TI-82

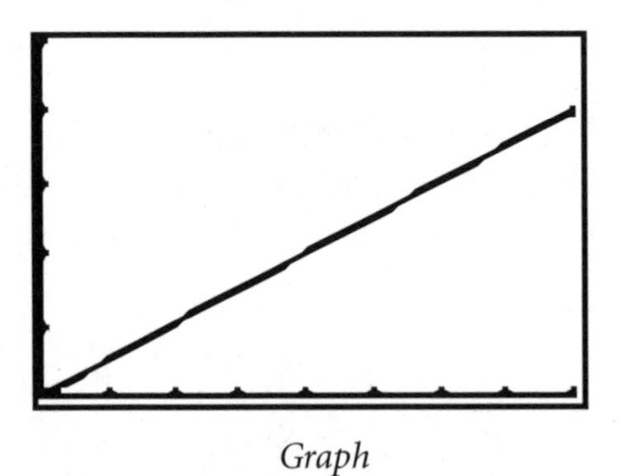

Graph

From the equations, it can be seen that this is the graph of $y = \frac{1}{2}x$. If the person had started at (1, 3) but moved at the same rate, the equations would be $x = 1 + 4t$ and $y = 3 + 2t$. The graph would be

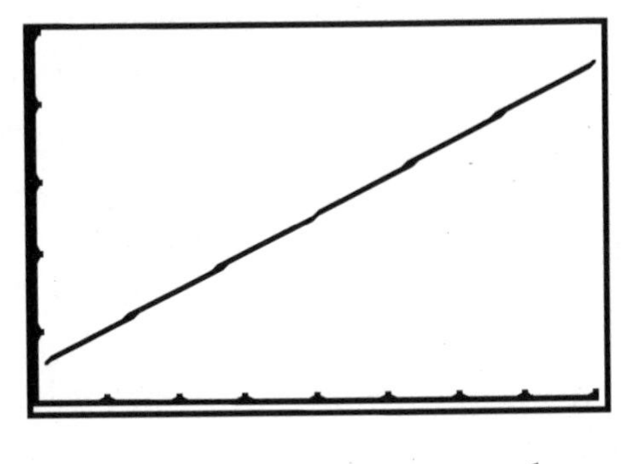

The equation of this line is $y = \frac{1}{2}(x+5)$.

Other Examples

Explore these:

$x = 4t - 10$	$x = t^2$	$x = 1 - t^2/2$	$x = 2\cos t$
$y = 1 - 3t$	$y = 1 - 3t$	$y = t - t^3/6$	$y = 5\sin t$
		$-5 \le t \le 5$	$-6.3 \le t \le 6.3$
		$-10 \le x \le 10$	(radian mode)
		$-7 \le y \le 7$	[–8,8] by [–6,6]

Sometimes it is easy to find y as a function of x; often it is not.

Starting with y = f(x) and Obtaining a Parametric Graph

When students see how to obtain the graph of $y = f(x)$ in parametric mode, they will understand the next step, which is to obtain a parametric graph from the equation $x = g(y)$.

Let $X_{1T} = T$; $Y_{1T} = f(T)$.

For example, if $y = x$^2 – 3x + 7, let $X_{1T} = T$; $Y_{1T} = T$^2 – 3T + 7.

To Graph x = g(y)

In this case, the roles of x and y are interchanged.

Let $X_{1T} = g(T)$; $Y_{1T} = T$

For example, if $x = y$^2, let $X_{1T} = T$^2; $Y_{1T} = T$.

Other Examples

1. To find the inverse function of $y = f(x)$, solve the equation $x = f(y)$ for y. To obtain the graph of the inverse function of $y = \frac{2+x}{3-x}$, graph $x = \frac{2+y}{3-y}$.

2. Graph the function $x = \sqrt{y^2 - 7y}$. Adjust the X-range, Y-range, and T-range until you see a complete graph (if the equation is squared, the underlying hyperbola is evident).

Computer Geometry Programs and the Experiments

The development of *interactive* or *dynamic* computer geometry programs is one of the most exciting—mathematically and pedagogically—new possibilities in teaching and learning geometry. More and different geometry can be studied, and students can construct their knowledge as they use the programs. The opportunities for explorations are endless. On the other hand, these programs are really electronic tools—compass and straightedge—equipped with memory and impressive calculating abilities. There are meaningful uses of these programs with the experiments, but their use should follow, not replace, the physical experiment.

When done using the physical equipment, the experiments are accessible to geometry students at every level. Using familiar apparatus—string, boards, graph paper—and familiar concepts—square, isosceles triangle, parallel—all students can perform the physical experiment and gather the data. Using basic ideas from geometry and algebraic manipulation, students can find the underlying function. The Interpret Your Findings questions enable them to study the function and the geometric relationship.

Replicating the experiment on the computer changes the focus. The geometry programs challenge the familiarity of the everyday concepts used in describing an experiment. Dividing a length of string into three segments of roughly the same size is easy. Dividing a line segment into thirds is a complex construction. There is a significant difference between physically shaping a loop of thread into the form of a rectangle and constructing a rectangle from a line segment of fixed length with electronic tools. Constructions involve parallel lines, bisections, and intersecting circles. Making the construction *dynamic,* so that the independent variable x can be changed by moving a point, is even more difficult. The focus of the experiment changes from *why* to *how.* Instead of asking, "Why do the data points lie on that particular curve?" the question becomes, "How do I keep the perimeter constant as I stretch the base?"

Simulating an experiment is a challenging extension. With a goal in mind—such as making the pet leash play out as the pet moves away—students have a reason to use the standard constructions of geometry. As students try to build a model of their physical activity, they will become aware of the fundamental connections between geometric figures and constructions. Replicating the physical activity of an experiment is an end in itself. The replication need not include replicating the measurements.

Some of the experiments have suggestions for dynamic extensions for which continued physical measurement is not as important as investigating the results. For these extensions, it makes sense to implement the entire experiment, including measurements, on a computer using software created for geometry explorations.

Teaching the Experiments

Whole-class Introduction

Use either Experiment 1, Perimeter of a Rectangle, or Experiment 2, Area of an Equilateral Triangle, as a whole-class introduction to the experiments. In both experiments, each student or group of students contributes one point to a common collection of data.

Model the procedure of plotting points on the calculator, as well as the processes on the Mathematical Analysis sheet, at an overheard projector. Students echo and extend the work at their desks. It is important that each student have a graphing calculator for this initial experiment. For later experiments, one calculator per group is adequate, but each group member should be able to enter and manipulate the data and expressions.

Subsequent Experiments

Before distributing the worksheets and equipment, discuss the nature of the experiment with the class. Make sure everyone knows what the independent and dependent variables are and whether the dependent variable can be measured directly or must be calculated.

Let students do as much on their own as possible. If a loop of thread is needed, let students cut and tie it. This will encourage ownership of the experiment and involvement in the results.

For all experiments, students should work in groups of two or three to collect data. Some experiments demand three people: two to hold objects and one to record. For others, two is ideal. *Each student should have a complete set of student sheets.* Although members of a group will confer during the analysis, *each is responsible for completing an individual sheet.* Similarly, each student should consider the Implications questions. Often these will give rise to discussions within the group.

Organizing and Analyzing Class Results

During the class period following an experiment, collect and display the equations at the end of each group's Mathematical Analysis. Discuss similarities and differences in the equations and graphs. Where appropriate, make a table of the maximum values (or intercepts) and ask, "Is there something going on that we could find out about?"

This is also a time to discuss extensions and variations on the experiment. In-class discussions can concentrate not only on the similarities and differences in the equations and graphs, but on the similarities and differences with the experimental variables. For example, Experiment 2, Area of an Equilateral Triangle, produces the equation $y = \frac{\sqrt{3}x^2}{4}$ and the graph:

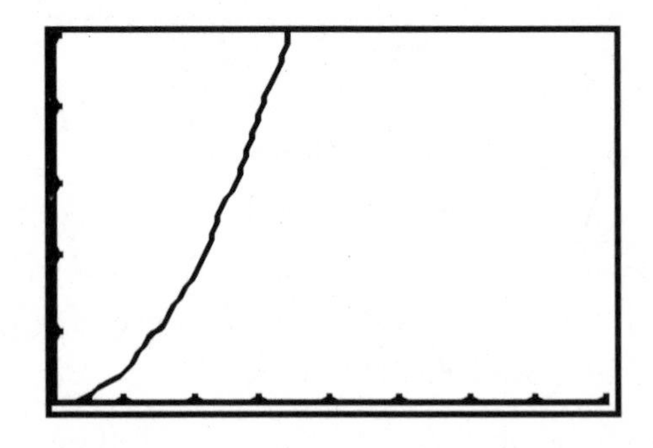

Ask students to substitute a square for the equilateral triangle and then to determine the equation and graph. Have other groups of students use regular pentagons, hexagons, or octagons. As the number of sides increases, ask students what they predict for n sides. Do the parabolas continue to get steeper, or do they approach a limit? (If the hardware and software are available, this makes an excellent extension for demonstration.) Talk about the units and what effect they have on the equation and the graph.

Assessing Each Student

Collect completed sheets from each student. Read them, make comments, and return the sheets to the students. The section on Assessment, which begins on page 8, has suggestions for other evaluative strategies. Because of their varied nature, the *Geometry Experiments* worksheets are especially suitable for inclusion in a student's portfolio, for class presentations, and for projects. They demonstrate writing, problem solving, cooperative learning, higher-order thinking skills, and the ability to make connections in mathematics and the use of technology. If worksheets are included in a student's portfolio, make a copy for the student to keep.

Index to the Experiments

	Experiment	Independent Variable	Dependent Variable
1	*Perimeter of a Rectangle*	base	perimeter
2	*Area of an Equilateral Triangle*	side	area
3	*Area of a Rectangle*	base	area
4	*Area of An Isoscles Triangle*	base	area
5	*Ramp Height*	distance from wall	height of ramp
6	*Pet Leash*	distance from hand	length of leash
7	*Casting Shadows*	distance from light source	length of shadow
8	*Stretching Points*	length of original line	length of stretched line
9	*Little House*	base of house	area of facade
10	*Reflections*	number of sides	mirror angle
11	*Looking Down*	distance from wall	ruler mark in mirror
12	*Polygons*	number of sides	sum of interior angles
13	*Stars*	number of sides	sum of the vertex angles
14	*Stretchy Isosceles Triangles*	height	distances between knots
15	*School Flower Beds*	width	area of flower bed
16	*Scaling the Wall*	ramp length	contact height
17	*Class Photo*	distance from end line	distance from camera
18	*Sliding Down*	distance from wall	distance from corner
19	*Squashed Boxes I*	base angle	area of the parallelogram
20	*Squashed Boxes II*	base angle	length of opposite diagonal

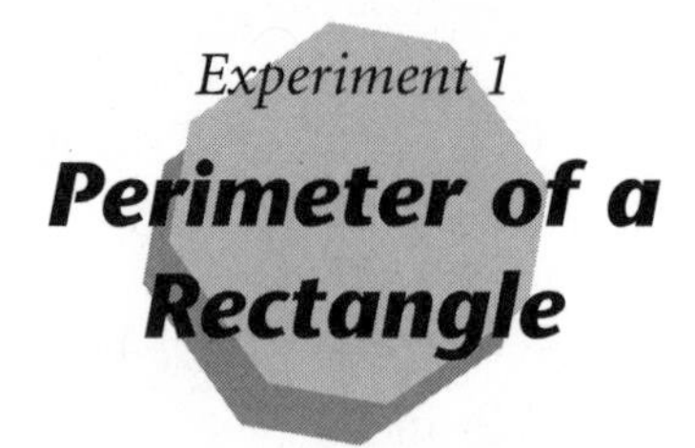

Teaching Notes

This experiment is a simple introduction to the process of looking for relationships between variables. The class works together to generate the data and to apply their geometric and algebraic knowledge to derive the algebraic relationship between the independent and dependent variables. In this experiment, the perimeter of a rectangle of (approximately) fixed area is a function of the base of the rectangle. The length of the base is the *independent variable,* and the perimeter of the rectangle is the *dependent variable.*

Key Concepts

Perimeter of a rectangle = 2 × (base + height)

Area of a rectangle = base × height

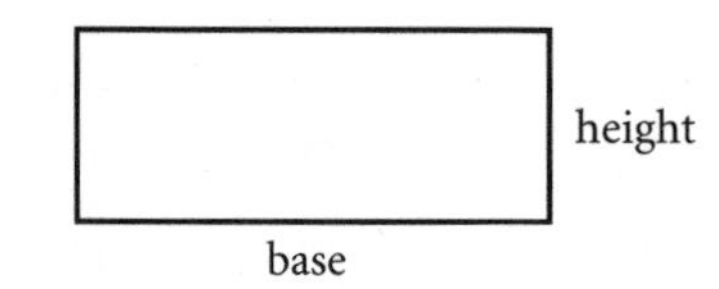

Equipment

graph paper (see page 131), 1 sheet per student

transparency of the Collect the Data sheet

overhead projector

overhead projector graphing calculator

Rectangles Sheet (see page 30), 1 per group

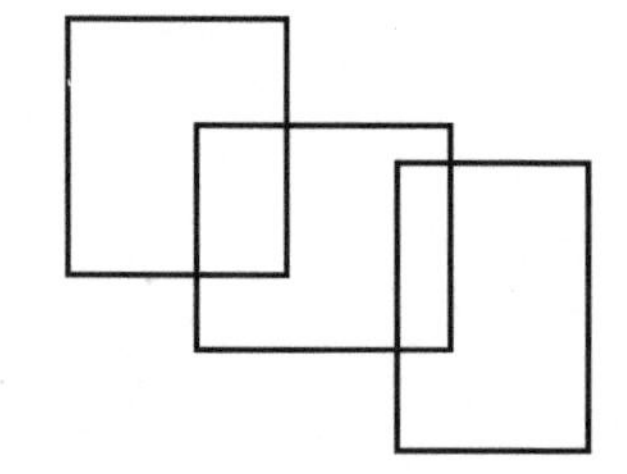

Procedure

Use this experiment to introduce the experiments to the entire class, to present or review how to use the graphing calculator, and to describe the process of the experiments with special attention to the Mathematical Analysis sheet. Students using the TI-82 should be familiar with the List and StatPlot menus; those using the TI-81 should be familiar with the Draw and Statistical menus.

In this experiment, students will be investigating the perimeter of a rectangle with an area of approximately 55 square units as a function of the base of the rectangle. However, students will gather data from rectangles with areas of 55 ± 1. The Rectangles Sheet provides a collection of rectangles with areas of 54 to 56 square units. Have each student choose a rectangle from the Rectangles Sheet, measure its base and height, then calculate its perimeter.

Call on students randomly to provide a data point until you have collected data from seven different rectangles. Enter the coordinates of the points in the x and y columns on the Collect the Data transparency; students should do the same on their Collect the Data sheets.

Before plotting the points, students must set the screen ranges. Xmin and Ymin are almost always best set to 0; set Xscl and Yscl to 0 also (no tic marks). For Xmax, choose a value somewhat larger (about half again) than the largest value of x in the data. Ymax should be larger than the largest value of y. For an area of 55, Ymax = 70 works well. Point out that it may be necessary to change Xmax and Ymax later to obtain a more complete view.

Students can obtain the Data Graph by using a statistical scatter plot or by writing a simple program that plots the points. Have students enter the points (x, y). When the coordinates of the last point have been entered, use the statistical scatter plot or run the program. The points will appear on the screen. Students should copy the locations of the dots as accurately as possible to the Data Graph on the Collect the Data sheet, allowing the points to remain on the calculator while they do the mathematical analysis.

Mathematical Analysis

If this is the first experiment your class is doing, work through the Mathematical Analysis sheet with students. Start with the rectangle of area

A and base x. Find the height, h, as a function of x and A: $h = \frac{A}{x}$.

Find the perimeter as a function of x and A: $P = 2\left(x + \frac{A}{x}\right)$.

Substitute the value of A (55) into this expression: $P = 2\left(x + \frac{55}{x}\right)$.

If this is correct, the graph of $y = 2\left(x + \frac{55}{x}\right)$ will pass *close* to the points.

On the TI-82, enter the function as **Y1** = $2(\text{X} + 55/\text{X})$ and press GRAPH.

On the TI-81, to avoid erasing the points from the screen, use the DrawF function: press 2nd PRGM 6 ENTER. The screen should be showing DrawF.

Type in the expression $2(\text{X} + 55/\text{X})$ and press ENTER.

The graph should be near or on most of the points. Because of the "errors," some points will still be visible.

On the transparency, sketch the graph of the function on top of the scatter graph. Point out that the experimental results and the theoretical analysis confirm each other.

Remind students that if the graph had completely missed the points, it would have been necessary to go back and find the mistake—which could have been in the analysis, the data collection, or the data entry.

Interpret Your Findings

Once the perimeter function has been found, there are several questions to be answered about the function and its graph. To analyze the graph on the TI-81, the function must be entered again as a Y-VAR. Enter it as **Y1** and press (GRAPH).

Inspect the graph on the screen: do the ranges need adjusting? If so, change Xmax and Ymax so that a more complete graph of the function is showing.

With the class, work through the questions on the Interpret Your Findings sheet. The first questions involve analyzing and interpreting the graph; they usually require the use of the (TRACE) and (ZOOM) functions. Use (TRACE) and (ZOOM) to find the minimum value of the perimeter. What does this rectangle look like?

Ask, "Is the value of the perimeter ever equal to the value of the area?" Demonstrate by setting **Y2** = 55 and pressing (GRAPH) to see both functions. Use (TRACE) and (ZOOM) to find where they intersect.

Some of the follow-up questions involve changing the equation to reflect a change in the experiment, then manipulating and/or graphing the equation to obtain the answer.

Experiment 1

Perimeter of a Rectangle

Name ______________________________

Partner(s) ______________________________

Key Concepts

Perimeter of a rectangle = 2 × (base + height)

Area of a rectangle = base × height

height

base

Collect the Data

Describe the experiment.

____

____

____

____

____

Data Collection

Base	Height	Perimeter

Points To Be Plotted

x Base	y Perimeter

Enter the points as data points in your calculator, then plot them. Copy the points from the calculator display to the screen diagram below. Record the screen ranges, and label the axes.

Data Graph

Xmin = ______________

Xmax = ______________

Ymin = ______________

Ymax = ______________

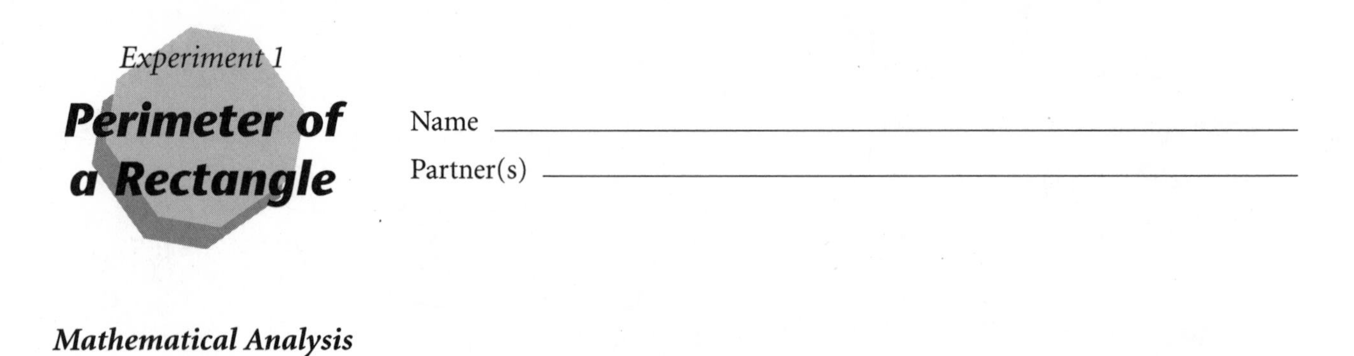

Mathematical Analysis

h

x

Write these equations for a rectangle with area A and base x.

Height = h = ______________________________
(in terms of x and A)

Perimeter = ______________________________
(in terms of x and h)

Substitute your expression for h:

Perimeter = ______________________________
(in terms of x and A)

What is the approximate area of each of your rectangles? ____________

For your rectangles, what is the area function (substitute your value of A)?

Perimeter = y = ______________________________

Graph your function (if you have a TI-82, use (Y=) and (GRAPH); if you have a TI-81, use DrawF). Do your data points lie *close* to the graph? If not, go back and determine the source of your error. Add the graph of the function to the Data Graph on the Collect the Data sheet.

Experiment 1

Perimeter of a Rectangle

Name ______________________________

Partner(s) ______________________________

Answer the following questions. Show your work. If you have not entered the function as **Y1**, do so now and graph it.

Interpret Your Findings

1. Use (TRACE) and (ZOOM) to find the minimum possible value of the perimeter.

   The minimum perimeter = ______________________

   It occurs when base = ______________________

2. For the perimeter and base in question 1, determine the height. ______

   What is special about this rectangle? ______________

3. Investigate whether the numerical value of the perimeter and the area are ever the same. To help find this value, set Y**2** = 55. Graph both functions. Use (TRACE) and (ZOOM) to find out where they intersect (if they do). For each value of x, find the height:

   Perimeter = area when x = __________; height = __________

   Perimeter = area when x = __________; height = __________

4. If the area had been 10 units smaller, what would the minimum perimeter have been? ______________________

   Explain in detail how you obtained your answer.

   ____

   ____

   ____

   ____

   ____

Rectangles Sheet

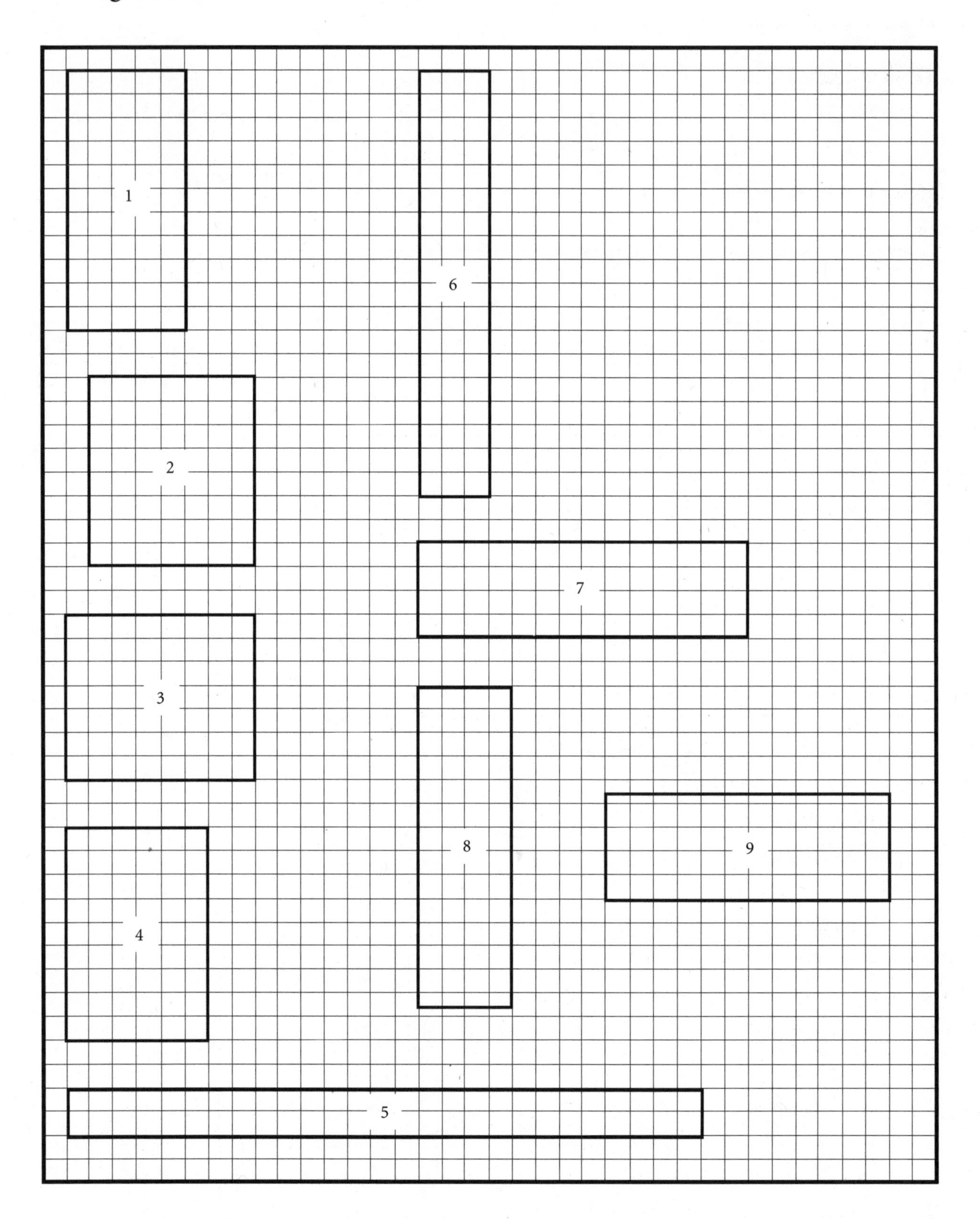

Experiment 2

Area of an Equilateral Triangle

Teaching Notes

In this experiment, students investigate the area of an equilateral triangle as a function of side length. The class generates a common data set. Each student contributes one data point by making an equilateral triangle from a loop of thread, then finding the triangle's area from the base and height. The length of the side is the *independent variable,* and the area of the triangle is the *dependent variable.*

Key Concepts

Pythagorean Theorem: $a^2 + b^2 = c^2$

Area of a triangle $= \frac{1}{2}$ base $\times$ height

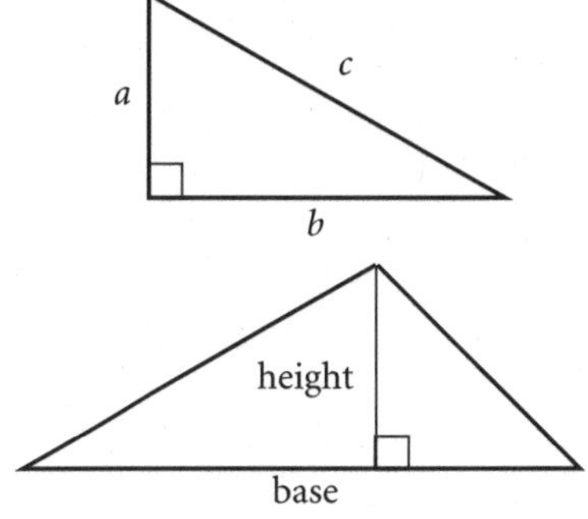

Equipment

graph paper (see page 131) with cardboard backing, 1 sheet per group

lengths of heavy, colored thread or string, 8"–20", 1 per group

Tie the ends to make a loop.

pushpins (to anchor the vertices of the triangle), 4 per group

transparency of the graph paper

overhead projector

Whole-class Introduction to the Experiments

Use this experiment to introduce the experiments to the entire class, to present or review how to use the graphing calculator, and to describe the process of the experiments with special attention to the Mathematical Analysis sheet. Students using the TI-82 should be familiar with the List and StatPlot menus; those using the TI-81 should be familiar with the Draw and Statistical menus.

Procedure

Demonstrate the procedure at the overhead projector with a loop of thread on the graph-paper transparency. Form what looks like an equilateral triangle, measure its base and height to the nearest half or quarter square, then calculate its area.

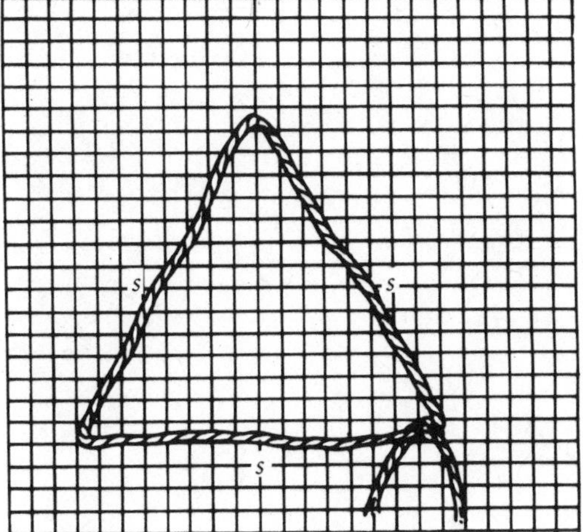

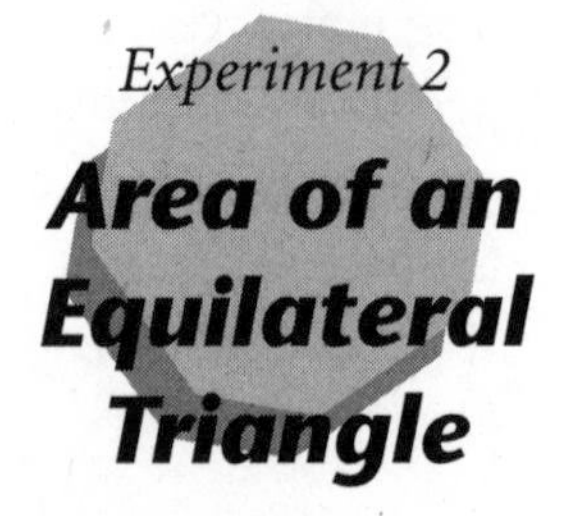

Teaching Notes, page 2

Collect data—length of side, height, and area—from each of the students. Select six different students to contribute points to use as the common data set. If a student's point has not been selected, that point will be the seventh point in his or her data set. Students whose points have already been selected should pick one more point to use as their seventh.

Have each student record the common data, augmented with their seventh point, on the Collect the Data sheet and use it to plot the points and complete the Mathematical Analysis and Interpret Your Findings sheets.

Mathematical Analysis

If this is the first experiment your class is doing, go over the formulas and complete the Mathematical Analysis sheet together.

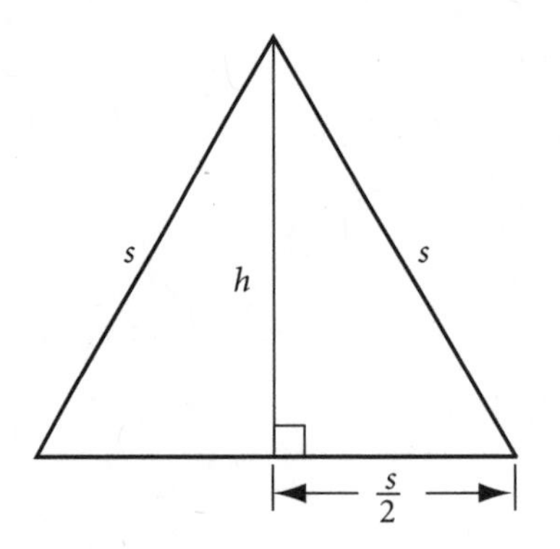

If the side is s and the height is h, the area is given by the formula

$\text{Area} = \frac{1}{2}sh.$

Using the Pythagorean Theorem, we get $h^2 + \left(\frac{s}{2}\right)^2 = s^2$.

Solving this formula for h gives $h = \frac{s}{2}\sqrt{3}$.

Substituting this value into the area formula gives Area = $y = \frac{s^2}{4}\sqrt{3}$.

Discuss appropriate values for Xmax and Ymax. These will depend on the lengths of the loops and the units on the graph paper. If needed, demonstrate how to enter data and to plot points.

Experiment 2

Area of an Equilateral Triangle

Name ______________________________

Partner(s) ______________________________

Key Concepts

Pythagorean Theorem: $a^2 + b^2 = c^2$

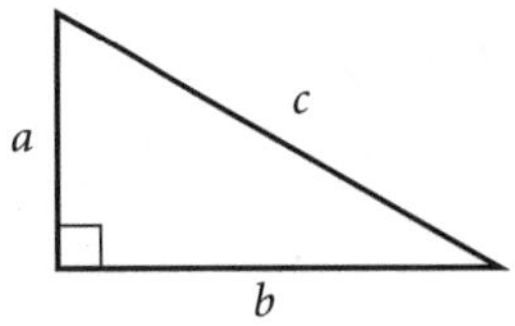

a
c
b

Collect the Data

Area of a triangle $= \frac{1}{2}$ base $\times$ height

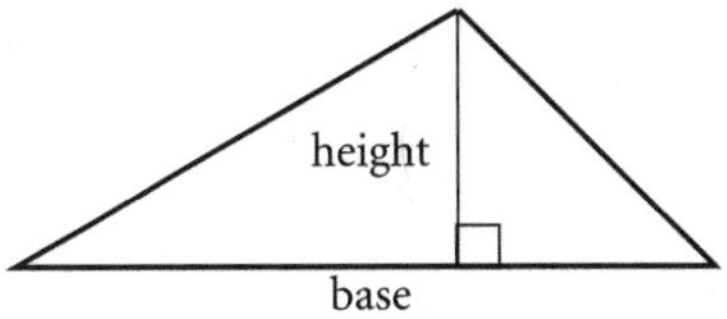

height
base

Describe the experiment.

____

____

____

____

Data Collection

Side	Height	Area

Points To Be Plotted

x Side	y Area

Enter the points as data points in your calculator, then plot them. Copy the points from the calculator display to the screen diagram below. Record the screen ranges, and label the axes.

Data Graph

Xmin = ____________

Xmax = ____________

Ymin = ____________

Ymax = ____________

Experiment 2

Area of an Equilateral Triangle

Name ______________________________

Partner(s) ______________________________

Mathematical Analysis

s s h s

Use the Pythagorean Theorem to write an equation using h and s.

____

Solve your equation for h^2:

$h^2 =$ ______________________________

(in terms of s)

Height $= h =$ ______________________________

Now find the area of the triangle:

Area = ______________________________

(in terms of s and h)

Now substitute your expression for h.

Area = ______________________________

(in terms of s)

For all equilateral triangles, what is the area function (substitute x for s)?

Area $= y =$ ______________________________

(in terms of x)

Graph your function (if you have a TI-82, use (Y=) and (GRAPH); if you have a TI-81, use DrawF). Do your data points lie *close* to the graph? If not, go back and determine the source of your error. Add the graph of the function to your Data Graph on the Collect the Data sheet.

Experiment 2

Area of an Equilateral Triangle

Name ____

Partner(s) ______________________________________

Answer the following questions. Show your work. If you have not entered the function as **Y1**, do so now and graph it.

Interpret Your Findings

1. If area = 10, then $s =$ _______ .

2. If area = 20, then $s =$ _______ .

3. If area = 40, then $s =$ _______ .

4. Ignoring the difference in units of measurement, is it possible for the *value* of the area to be equal to the value of a side. Redraw the graph of **Y1**, and add the graph of **Y2** $= x$ to help you find two such points.

   Side = _______ Area = _______

   Side = _______ Area = _______

5. For *values* of the sides between the two points you found in question 4, which of the following is true?

   Side < Area Side > Area

6. Suppose two students used graph paper with smaller squares than your graph paper to measure their base and height. Would their points lie on your graph? _______ Explain your answer.

   ____

   ____

   ____

   ____

7. If you double the length of the side of the equilateral triangle, does the area double? ____________

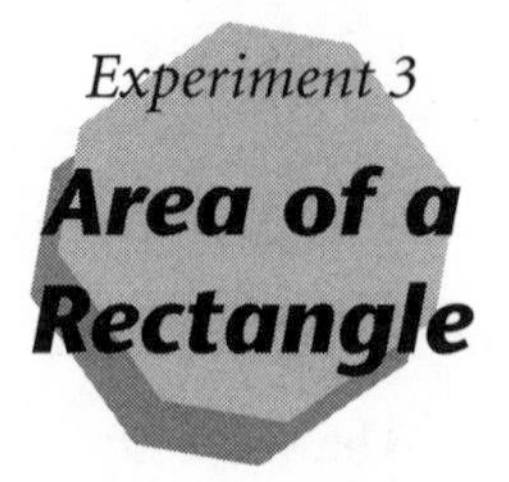

Teaching Notes

In this experiment, students find the area of a rectangle of fixed perimeter *P*. They will measure the base and height, then calculate the area. The base of the rectangle is the *independent variable,* and the area of the rectangle is the *dependent variable.*

Key Concepts

Area of a rectangle = base × height

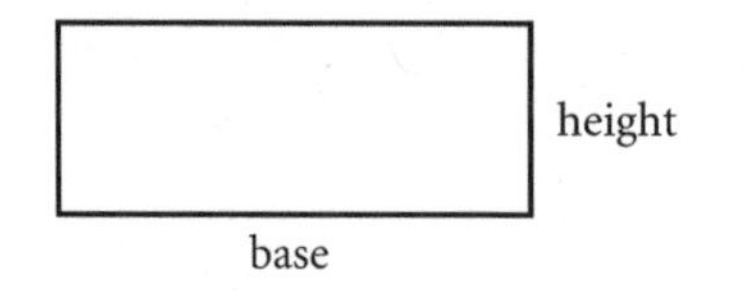

Perimeter of a rectangle = 2 × (base + height)

Equipment

graph paper (see page 131) with cardboard backing, 1 sheet per group

lengths of heavy, colored thread or string, 14"–20", 1 per group

Tie the ends to make a loop.

pushpins (to anchor the corners of the rectangle), 4 per group

transparency of the graph paper

overhead projector

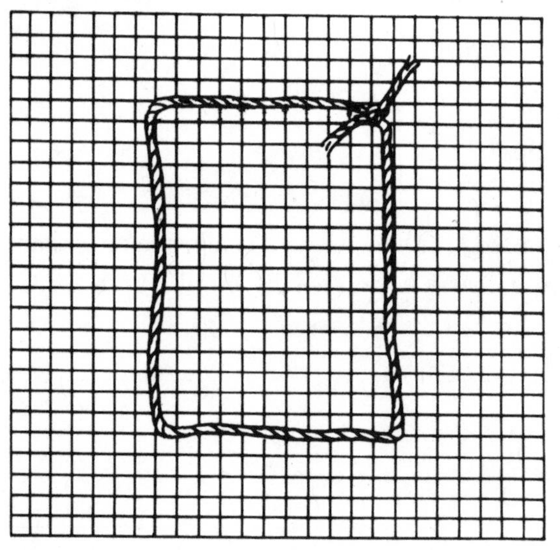

Procedure

Demonstrate the experiment at the overhead projector with a loop on the graph-paper transparency.

Students, working in groups of three, decide on the size of the base of a rectangle, using the squares on the graph paper as units. They form the rectangle with their thread, measure its height to the nearest half square or quarter square, then calculate the rectangle's area. They make and measure seven different rectangles. Encourage each group to include at least one short rectangle, one tall rectangle, and one approximately square rectangle.

Teaching Notes, page 2

After collecting their data, students set the screen ranges to show the first quadrant and plot the points (x, y), where y = area. They then find the equation of the area function and verify their analyses by graphing the function and observing how closely it fits the plotted points.

Organizing and Analyzing Class Results

From each group, collect the following information:

P	Area Function	Base of Largest Rectangle	Area of Largest Rectangle

For the rectangle of maximum area, ask students what the theoretical relationship is between the base and P and between the area and P. (If students do not immediately see that $b = \frac{P}{4}$, ask, "What was the shape of the largest rectangle?")

Have each group calculate $\frac{P}{4}$ and $\frac{P^2}{16}$. Have the class discuss the reasons for the differences between these values and their results in question 1 on the Interpret Your Findings sheet.

Mathematical Analysis

If the perimeter P is given by $P = 2(x + h)$, solving for h gives $h = \frac{P - 2x}{2}$.

Substituting this into the area formula $A = xh$ gives $A = y = x\left(\frac{P - 2x}{2}\right)$.

Experiment 3

Area of a Rectangle

Name ______________________________

Partner(s) ______________________________

Key Concepts

Area of a rectangle = base × height

height

base

Perimeter of a rectangle = 2 × (base + height)

Collect the Data

Describe the experiment.

____

____

____

____

____

Data Collection

Base	Height	Area

Points To Be Plotted

x Base	y Area

Enter the points as data points in your calculator, then plot them. Copy the points from the calculator display to the screen diagram below. Record the screen ranges, and label the axes.

Data Graph

Xmin = __________

Xmax = __________

Ymin = __________

Ymax = __________

Experiment 3

Area of a Rectangle

Name ______________________________

Partner(s) ______________________________

Mathematical Analysis

h

x

If a rectangle has perimeter P, write P in terms of x and h.

$P =$ ______________________________

Now solve for h.

Height $= h =$ ______________________________
(in terms of x and P)

Area = ______________________________
(in terms of x and h)

Now, substitute for h.
Area = ______________________________
(in terms of x and P)

What is the perimeter of each of your rectangles? ______________________________

How did you determine the value of P? ______________________________

____

____

For your rectangles, what is the area function (substitute your value of P)?

Area $= y =$ ______________________________

Graph your function (if you have a TI-82, use [Y=] and [GRAPH]; if you have a TI-81, use DrawF). Do your data points lie *close* to the graph? If not, go back and determine the source of your error. Add the graph of the function to the Data Graph on the Collect the Data sheet.

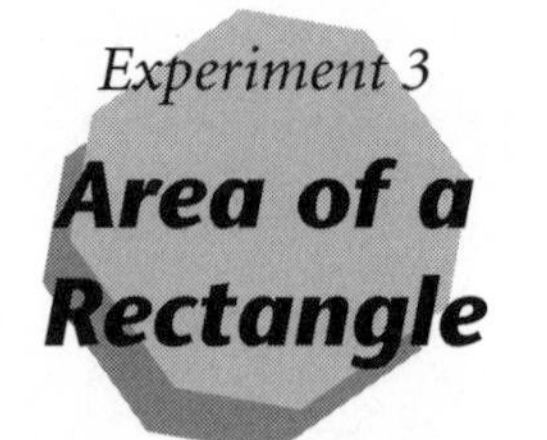

Area of a Rectangle

Name ______________________________

Partner(s) ______________________________

Answer the following questions. Show your work. If you have not entered the function as **Y1**, do so now and graph it.

Interpret Your Findings

1. Use (TRACE) and (ZOOM) to find the value of the nonzero x-intercept.

   ____

   Describe the physical significance of this point.

   ____

   ____

2. Use (TRACE) and (ZOOM) to find the maximum possible value of the area of a rectangle made with your loop:

   The maximum area = ________ . It occurs when the base = ________

3. Use your loop to make the rectangle of maximum area. What is special about it?

   ____

   ____

4. Had your perimeter been 3 units shorter, what would the maximum area have been? ________

   Explain in detail how you obtained your answer.

   ____

   ____

5. You did the Area of the Rectangle experiment and obtained the following graph. Using a different perimeter for her rectangles, Zia did the same experiment. Can you determine the perimeter of her rectangles? ________ Explain your method.

   ____

   ____

   ____

   ____

   ____

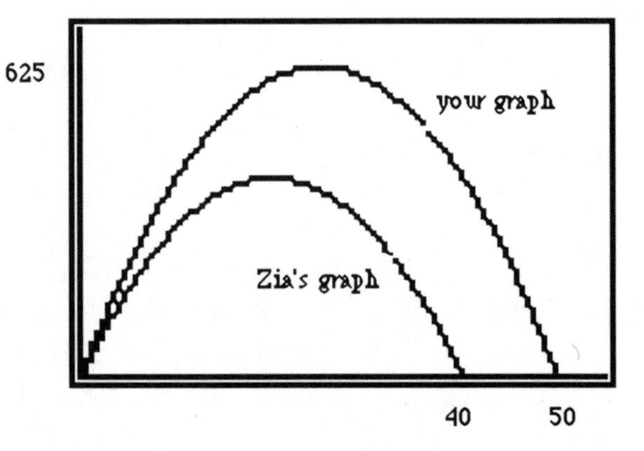

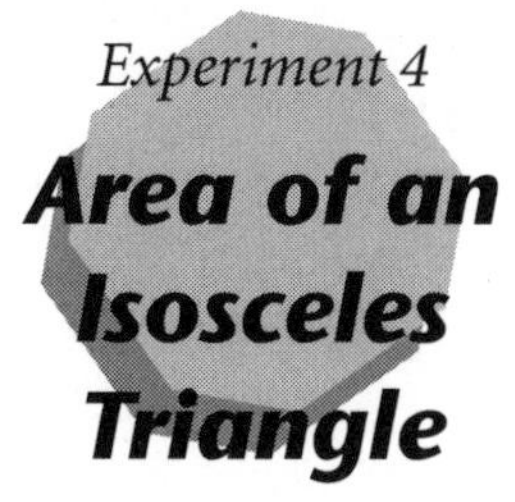

Experiment 4

Area of an Isosceles Triangle

Teaching Notes

In this experiment, students calculate the area of an isosceles triangle of fixed perimeter P after measuring its base and height. The *independent variable* is the base, x, of the isosceles triangle, and the *dependent variable* is the area of the isosceles triangle.

Key Concepts

Perimeter of a triangle = $a + b + c$

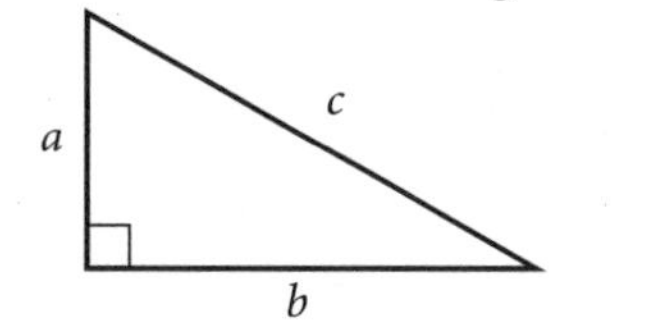

Pythagorean Theorem: $a^2 + b^2 = c^2$

Equipment

graph paper (see page 131) with cardboard backing, 1 sheet per group

lengths of heavy, colored thread or string, 14"–20", 1 per group

Tie the ends to make a loop.

pushpins to anchor the corners, 3 per group

transparency of the graph paper

overhead projector

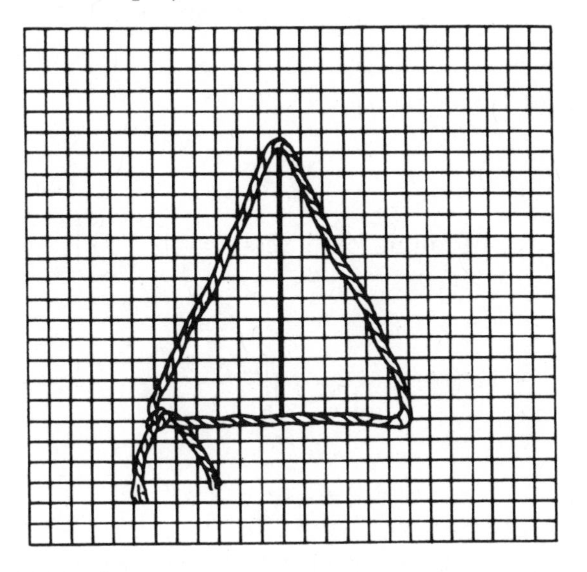

Procedure

Demonstrate at the overhead projector. Form the string loop into an isosceles triangle, with base of about 2", on the graph-paper transparency. If necessary, darken a vertical line in the center of the graph paper.

Using the squares on the graph paper as the units, students, working in groups of two or three, decide on the size of the base of an isosceles triangle. The center of the base should be at a thick line of the graph paper. Stretching the vertex along the thick line, students form an isosceles triangle, measure its height, then calculate its area. Encourage students to create and measure both short and tall isosceles triangles.

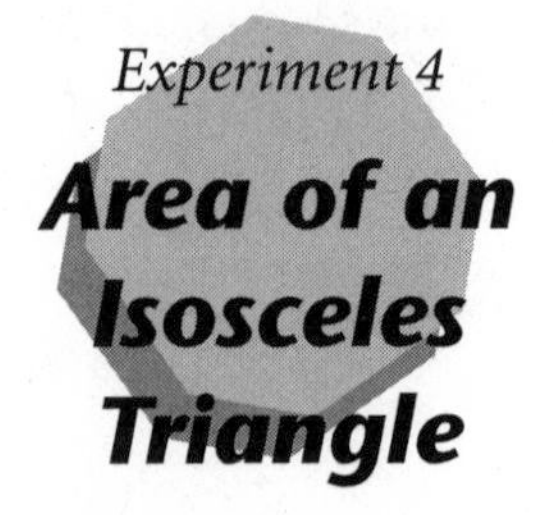

Experiment 4

Area of an Isosceles Triangle

Teaching Notes, page 2

Once they have measured six or seven isosceles triangles, students plot the points (x, y), where y = area. They will have to set appropriate screen ranges; the screen should show the first quadrant. They then find the equation of the area function and verify their analyses by graphing the function and observing how closely it fits the plotted points.

Mathematical Analysis

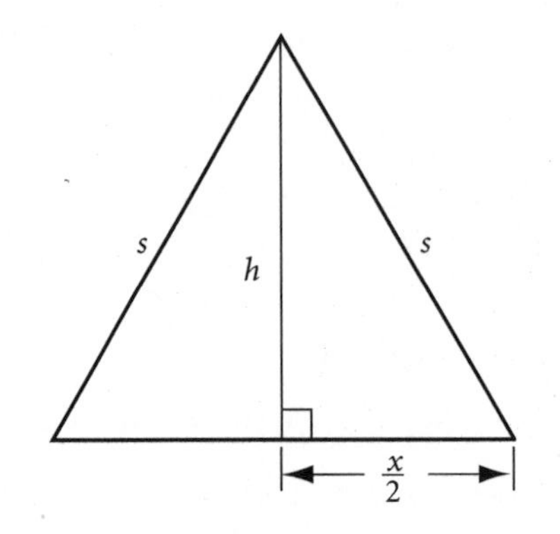

If the perimeter is P, then $P = 2s + x$ and $\dfrac{P-x}{2}$.

Using the Pythagorean Theorem, we get $\left(\dfrac{x}{2}\right)^2 + h^2 = s^2$.

Solving for h gives $h = \sqrt{s^2 - \left(\dfrac{x}{2}\right)^2}$.

Substituting for s gives $h = \sqrt{\left(\dfrac{P-x}{2}\right)^2 - \left(\dfrac{x}{2}\right)^2}$.

The area is $y = \dfrac{1}{2}x\sqrt{\left(\dfrac{P-x}{2}\right)^2 - \left(\dfrac{x}{2}\right)^2} = \dfrac{x}{4}\sqrt{P^2 - 2Px}.$

Extension

Students will find that of all isosceles triangles with fixed perimeter P, the triangle with the maximum area is an equilateral triangle. Ask students, "For a fixed perimeter P, is there a non-isosceles triangle with a greater area?" Have students experiment with their loops and record their conclusions.

Experiment 4

Area of an Isosceles Triangle

Collect the Data

Name ____________________

Partner(s) ____________________

Key Concepts

Perimeter of a triangle = $a + b + c$

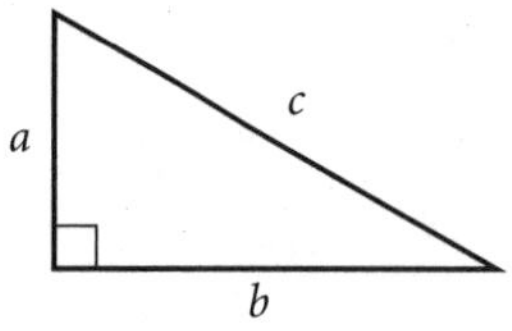

Pythagorean Theorem: $a^2 + b^2 = c^2$

Describe the experiment.

____

____

____

____

____

Data Collection

Base	Height	Area

Points To Be Plotted

x Base	y Area

Enter the points as data points in your calculator, then plot them. Copy the points from the calculator display to the screen diagram below. Record the screen ranges, and label the axes.

Data Graph

Xmin = ________

Xmax = ________

Ymin = ________

Ymax = ________

Experiment 4

Area of an Isosceles Triangle

Name ______________________________

Partner(s) ______________________________

Mathematical Analysis

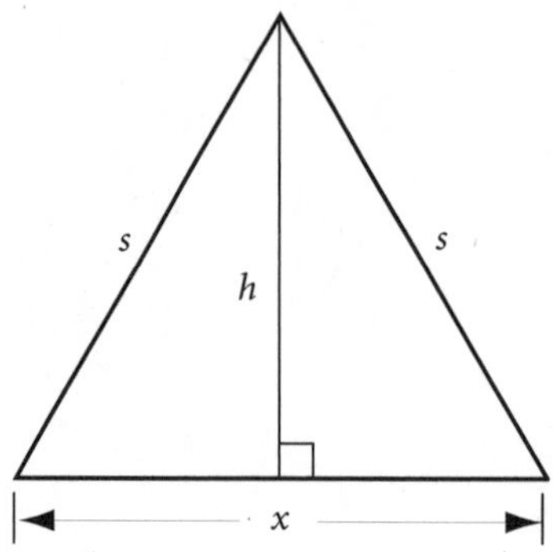

An isosceles triangle has perimeter P and base x.

Find P. $P =$ ______________________________
(in terms of x and s)

Solve for s. $s =$ ______________________________
(in terms of x and P)

$h^2 =$ ______________________________
(in terms of x and s)

Substitute for s: $h =$ ______________________________
(in terms of x and P)

Area = ______________________________
(in terms of x and P)

What is the perimeter of each of your isosceles triangles?______________

How did you determine the value of P?

____

____

Substitute your value of P in your area function.

Area = $y =$ ______________________________

Graph your function (if you have a TI-82, use (Y=) and (GRAPH); if you have a TI-81, use DrawF). Do your data points lie *close* to the graph? If not, go back and determine the source of your error. Add the graph of the function to the Data Graph on the Collect the Data sheet.

Experiment 4

Area of an Isosceles Triangle

Name ______________________________

Partner(s) ______________________________

Answer the following questions. Show your work. If you have not entered the function as **Y1**, do so now and graph it.

Interpret Your Findings

1. Use (TRACE) and (ZOOM) to find the value of the nonzero *x*-intercept.
   __________ What triangle corresponds to this value of *x*?

   ____

2. Use (TRACE) and (ZOOM) to find the maximum possible value of the area of an isosceles triangle made with your loop.
   The maximum area is __________ . It occurs when $x =$ __________ .

3. Use your loop to make the triangle of maximum area.
   What is its height? __________ What are the sides? __________
   What is special about this triangle?

   ____

4. What is the connection between the value of *x* in question 2 and *P*? (Use an equation.)

   ____

   ____

   ____

5. If your loop had been twice as long, would the maximum area have been double? ______________________________

   Explain in detail how you obtained your answer.

   ____

   ____

   ____

   ____

Teaching Notes

The height of a ramp of fixed length is a function of the distance from the foot of the ramp to the wall against which it leans. This experiment is an investigation based on the Pythagorean Theorem. The distance between the ramp and the wall is the *independent variable,* and the height of the ramp is the *dependent variable.*

Key Concept

Pythagorean Theorem: $a^2 + b^2 = c^2$

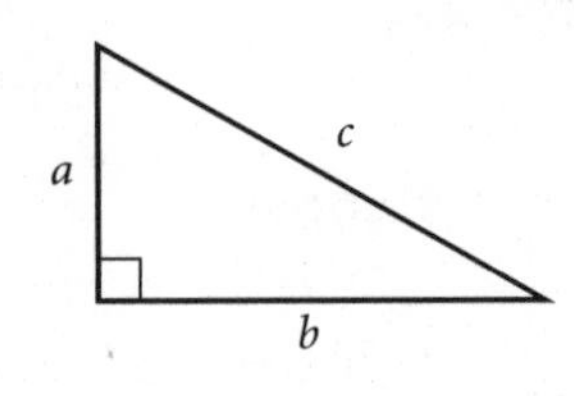

Equipment

boards of varying lengths (12"–24"), 1 board per group

yardsticks or meter sticks, 1 per group

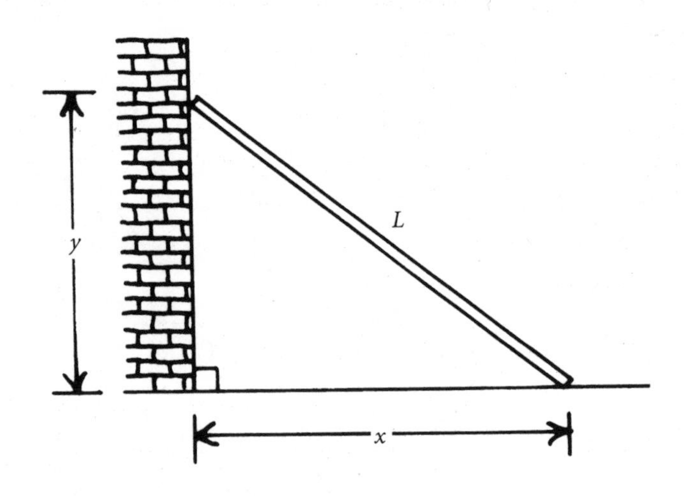

Procedure

Students, working in groups of two or three, lean the ramp against the wall. They measure the distance from the bottom of the ramp to the wall and record the value as the independent variable x. The dependent variable is the height of the ramp.

Depending on how the screen ranges are set, the points will appear to lie on either a circle or an ellipse.

The mathematical analysis involves manipulating the Pythagorean Theorem, $y^2 + x^2 = L^2$, to this form: $y' = \sqrt{L^2 - x^2}$.

Extension

You might ask students such questions as, "What value of x yields the maximum area?" "What kind of triangle is it?" "Why does the graph seem to just stop when $x > L$?"

Ask students what happens if the dependent variable is changed to be the area of the triangle. They can use the data they have already collected to investigate this situation.

Experiment 5

Ramp Height

Name ______________________________

Partner(s) ______________________________

Key Concept

Pythagorean Theorem: $a^2 + b^2 = c^2$

a

c

b

Collect the Data

Describe the experiment.

____

____

____

____

____

Data Collection

Distance to Wall	Height	Area

Points To Be Plotted

x	y

Enter the points as data points in your calculator, then plot them. Copy the points from the calculator display to the screen diagram below. Record the screen ranges, and label the axes.

Data Graph

Xmin = ____________

Xmax = ____________

Ymin = ____________

Ymax = ____________

Experiment 5

Ramp Height

Name ______________________________

Partner(s) ______________________________

Mathematical Analysis

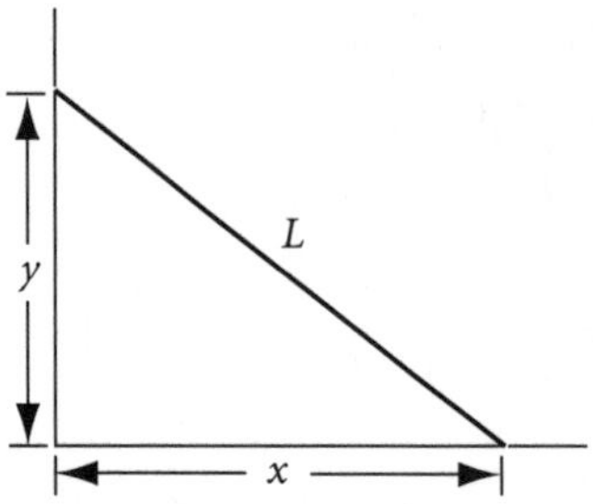

For a ramp of length L, the distance x, from the corner (base of the wall) and the height y, are related by the equation:

____

Give the equation for the ramp height, y.

Height = y = ______________________________

(in terms of x and L)

What is the length of your ramp? ______________________________

For your ramp, the height is given by what function?

Height = y = ______________________________

Graph your function (if you have a TI-82, use (Y=) and (GRAPH); if you have a TI-81, use DrawF). Do your data points lie *close* to the graph? If not, go back and determine the source of your error. Add the graph of the function to the Data Graph on the Collect the Data sheet.

Name ______________________________

Partner(s) ______________________________

Answer the following questions. Show your work. If you have not entered the function as **Y1**, do so now and graph it.

Interpret Your Findings

1. Use (TRACE) and (ZOOM) to find the value of the x-intercept. ________
   Describe the physical significance of this point. Draw a picture showing the position of the ramp corresponding to this data point.

   ____

2. Use (TRACE) and (ZOOM) to find the value of the y-intercept. ________
   Describe the physical significance of this point.

   ____

   ____

   ____

3. Had your ramp been 6 units shorter, what would the graph have looked like? What is the x-intercept? What is the equation for the height of the shorter ramp?

   ____

   ____

   ____

   Graph this equation. Add this graph to your "screen" on the Collect the Data sheet and label it.

4. Suppose, using the same ramp, that your partner measured in inches, while you measured in centimeters. Sketch and label the two height functions.

Teaching Notes

In this experiment, students measure the distance of a "pet" from a fixed point (at the owner's feet) and the length of the leash. The distance of the pet from the fixed point is the *independent variable,* and the length of the leash is the *dependent variable.*

Key Concept

Pythagorean Theorem: $a^2 + b^2 = c^2$

Equipment

"leashes" (4-foot lengths of string or cord), 1 per group

"pets" (small figures and peg people work well), 1 per group

yardsticks or meter sticks, 1 per group

Procedure

Have students tie one end of the "leash" to the "pet." The other end of the leash and all the slack is held in the owner's hand (which must remain in a fixed position). Tying the leash to a fixture is a way to simulate this situation.

Students then let the pet "wander" from its owner, who gradually lets out the leash, keeping it taut. As the pet walks away, students measure the length of the leash for different values of x.

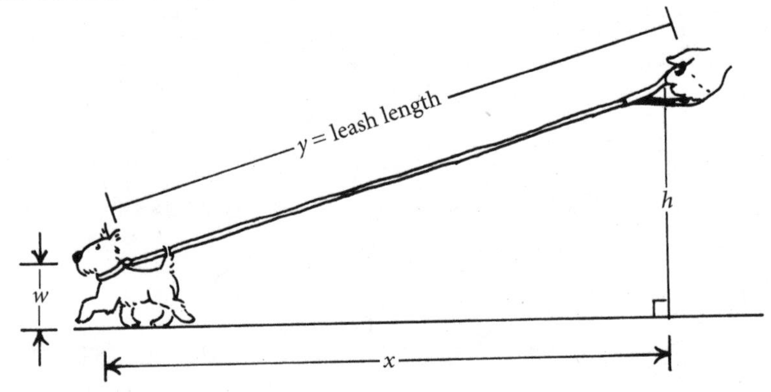

The height of the owner's hand, h, remains constant. Students may decide to reduce h by w when verifying their mathematical analyses.

Experimental Results

For values of $x < h$, the graph is nonlinear; as x increases, the function appears to be linear.

Mathematical Analysis

By the Pythagorean Theorem, the leash length = $y = \sqrt{(h-w)^2 + x^2}$.

If some students' analyses produce a curve of the correct form but lying below the data points, suggest that they consider the *real height, h,* of the triangle (that is, that they take w into account, where w is the distance from the ground to the pet's neck). For larger values of x, both $\sqrt{h^2 + x^2}$ and $\sqrt{(h-w)^2 + x^2}$ are very close to $\sqrt{x^2} = x$. Question 3 asks students to explain this asympotic behavior.

Experiment 6

Pet Leash

Name ______________________________

Partner(s) ______________________________

Key Concept

Pythagorean Theorem: $a^2 + b^2 = c^2$

a c b

Collect the Data

Describe the experiment.

____

____

____

____

____

Data Collection

x	Leash Length

Points To Be Plotted

x	*y*

Enter the points as data points in your calculator, then plot them. Copy the points from the calculator display to the screen diagram below. Record the screen ranges, and label the axes.

Data Graph

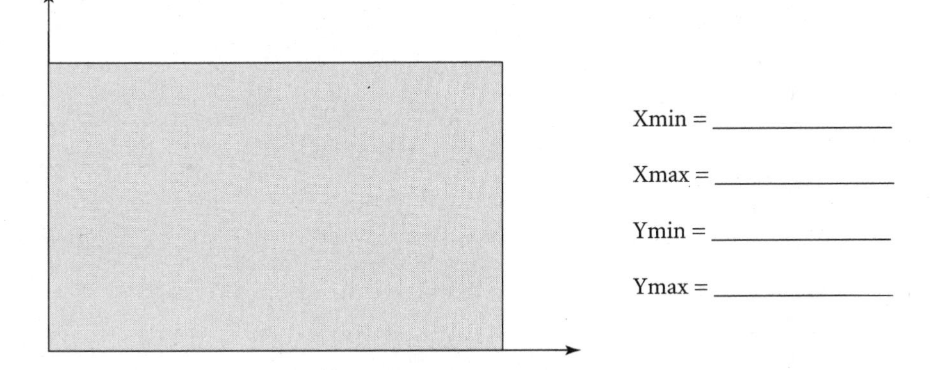

Name ______________________________

Partner(s) ______________________________

Mathematical Analysis

Assume the hand has height h.

Express the leash length y in terms of x and h.

$y =$ ______________________________
(in terms of x and h)

For your experiment, $h =$ ______________________________.

For your setup, what is the function (substitute your value of h)?

$y =$ ______________________________
(in terms of x)

Graph your function (if you have a TI-82, use Y= and GRAPH; if you have a TI-81, use DrawF). Do your data points lie *close* to the graph? If not, go back and determine the source of your error. Add the graph of the function to the Data Graph on the Collect the Data sheet.

Name ______________________________

Partner(s) ______________________________

Answer the following questions. Show your work. If you have not entered the function as **Y1**, do so now and graph it.

Interpret Your Findings

1. Use (TRACE) or algebra to find the value of x for which the length of the leash is twice the value of x. ________

2. Place the pet at the value of x found in question 1. Measure the angle the (extended) leash makes with the floor. ________

3. As x increases, the graph of the function appears to be a line. What is the equation of this line? Graph both the function and the line, and copy the graphs in the box.

Do the graphs ever meet? When? Or why not?

____

____

____

4. Change Xmin to the negative of your value for Xmax, and redraw the graph. Write a paragraph describing how this graph relates to the pet's wanderings.

____

____

____

____

Experiment 7

Casting Shadows

Teaching Notes

In this experiment, students measure the length of the shadow of a "doll" and its distance from the light source. The *independent variable* is the doll's distance from the "lamppost," and the *dependent variable* is the length of the shadow.

Key Concept

Similar triangles: Corresponding sides are proportional

Equipment

light sources, 1 per group

A utility lamp hung 3–4 feet from the floor will work, or a secured flashlight.

yardsticks or meter sticks, 2 per group

standing figures, 3"–8" tall, 1 per group

Procedure

Have students tape their yardsticks to the floor. Students "walk" the doll along the yardstick, measuring the shadow length for different values of x. Since shadows grow slowly, students must measure precisely (to the nearest half centimeter or the nearest eighth of an inch). If the shadow is not sharp, they can place a sheet of white paper alongside the yardstick.

The height of the doll, d, and the height of the lamp, h, affect the value of y.

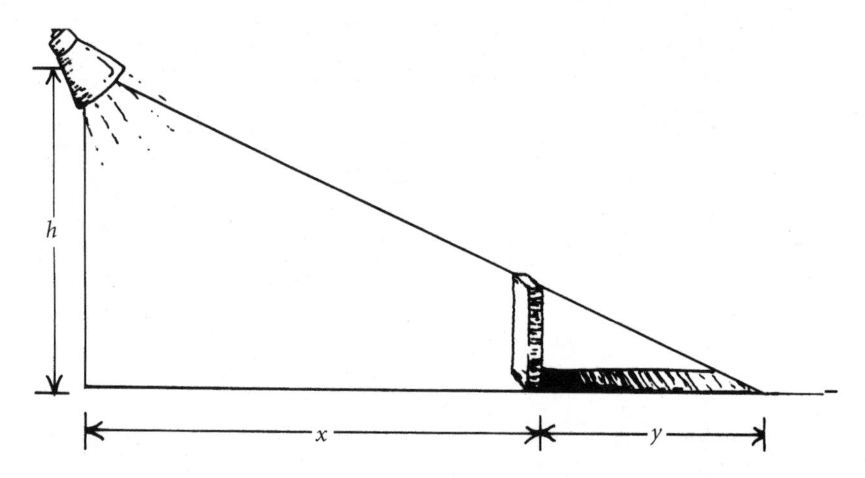

Teaching Notes, page 2

Mathematical Analysis

The data points should be linear. If students use similar triangles and the ratio $\frac{d}{h} = \frac{y}{x+y}$ and solve for y, they will get $y = \frac{dx}{h-d}$.

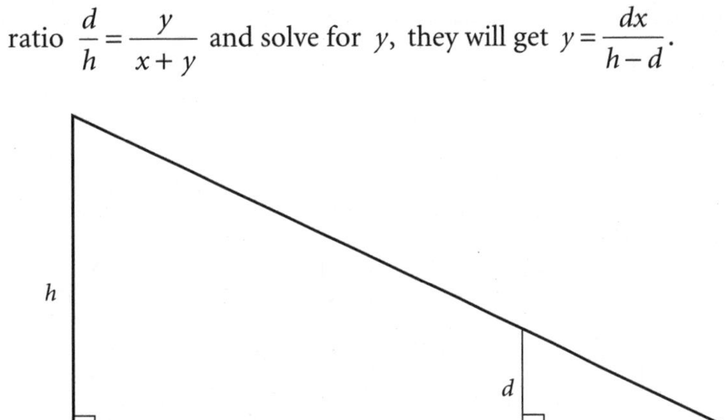

The graph of the function is very sensitive to the difference of $h - d$. Students may find their line is parallel to, but not exactly on top of, the line implied by the data points.

Experiment 7

Casting Shadows

Name ____________________

Partner(s) ____________________

Key Concept

Similar triangles: Corresponding sides are proportional

Collect the Data

Describe the experiment.

____

____

____

____

____

____

Units used: ________

Data Collection

x	Shadow Length

Points To Be Plotted

x	y

Enter the points as data points in your calculator, then plot them. Copy the points from the calculator display to the screen diagram below. Record the screen ranges, and label the axes.

Data Graph

Xmin = ________

Xmax = ________

Ymin = ________

Ymax = ________

Experiment 7

Casting Shadows

Name ______________________________

Partner(s) ______________________________

Mathematical Analysis

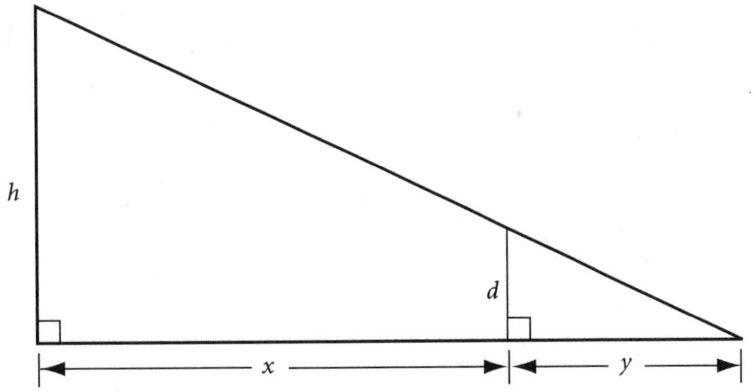

h

d

x

y

Assume the light source has a height of h and the doll has a height of d. Using corresponding sides of similar triangles, complete the equation below.

$\frac{d}{h} =$ ______________________________

Solve for y.

$y =$ ______________________________
(in terms of x, h, and d)

For your experiment, what are h and d?

$h =$ __________; $d =$ __________

For your setup, what is the function (substitute your values of h and d)?

$y =$ ______________________________
(in terms of x)

Graph your function (if you have a TI-82, use (Y=) and (GRAPH); if you have a TI-81, use DrawF). Do your data points lie *close* to the graph? If not, go back and determine the source of your error. Add the graph of the function to the Data Graph on the Collect the Data sheet.

Name ______________________________

Partner(s) ______________________________

Answer the following questions. Show your work. If you have not entered the function as **Y1**, do so now and graph it.

Interpret Your Findings

1. Use (TRACE) and (ZOOM) or algebra to find the smallest value of x for which the length of the shadow is the same as the height of the doll.

   ____

2. For what value of x will the length of the shadow be twice the height of the doll? ________

3. Suppose the doll had been twice as tall. Would the shadow have been exactly twice as long? More than twice as long? Less? Does it depend on x? Explain your reasoning.

   ____

   ____

   ____

4. What do you think would happen if the lamp were the same height as the doll?

   ____

   Does your equation predict this? ________ How?

   ____

   ____

5. Suppose another group's line was steeper than the line from your group. Assume both groups collected the data and solved the equations correctly. What factors in the experiment could account for the difference? Explain.

   ____

   ____

   ____

Teaching Notes

In this experiment, students measure the distance of a knot in a rubber band to a base line and the distance from the end of the rubber band to the base line. The length of the vertical line segment to the knot is the *independent variable,* and the length of the vertical line segment at the end of the rubber band is the *dependent variable.*

Key Concept

Similar triangles: Corresponding sides are proportional

Equipment

graph paper (see page 131) with cardboard backing, 1 sheet per group

long rubber bands, 1 per group

transparency of the graph paper

overhead projector

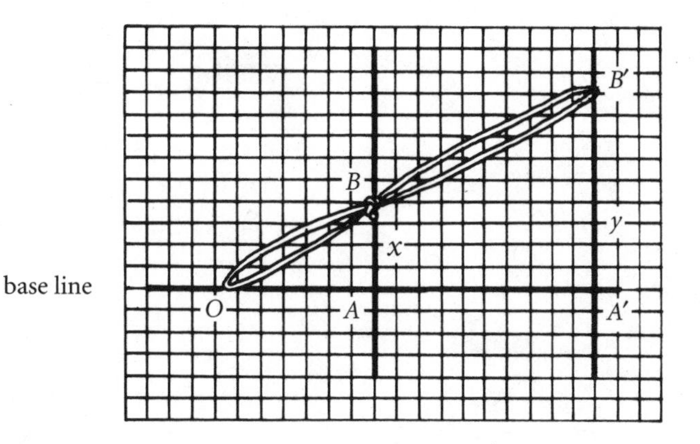

Procedure

Demonstrate the following procedure at the overhead projector:

Knot one of the long rubber bands off-center. Draw a base line, L, on the graph-paper transparency, and mark the point A. If necessary, darken the vertical line through A. The anchor point, O, lies on L and should not be at the edge of the page. When one of the ends of the rubber band is on O and the knot is on A, the band should be stretched a bit. Anchor the band at O with your finger or a pencil, then stretch it until the knot lies on the line through A. Measure x and y, where y is the distance from the stretched end of the band from the line L.

As students work, they will discover that the distances, y, are all measured on the same vertical line.

Teaching Notes, page 2

Mathematical Analysis

Students will discover that OA and OA' are constant for the experiment.

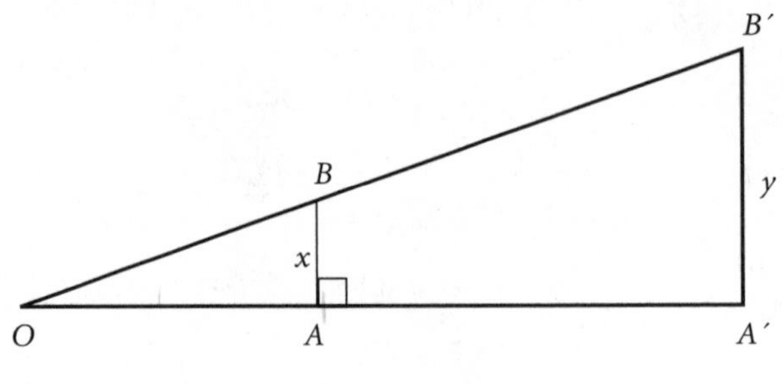

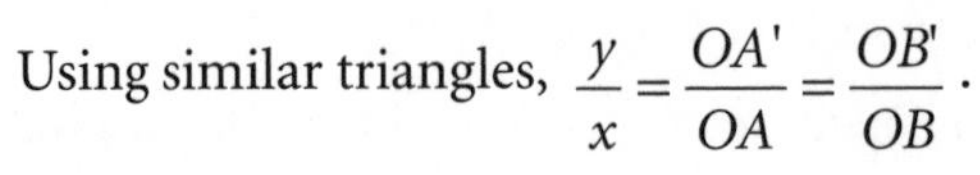

Using similar triangles, $\frac{y}{x} = \frac{OA'}{OA} = \frac{OB'}{OB}$.

$$y = \frac{OA'}{OA}x = Kx .$$

Experiment 8

Stretching Points

Name ______________________________

Partner(s) ______________________________

Key Concept

Similar triangles: Corresponding sides are proportional

Collect the Data

Describe the experiment.

____

____

____

____

____

____

Data Collection

AB	$A'B'$

Points To Be Plotted

x	y

Enter the points as data points in your calculator, then plot them. Copy the points from the calculator display to the screen diagram below. Record the screen ranges, and label the axes.

Data Graph

Xmin = ____________

Xmax = ____________

Ymin = ____________

Ymax = ____________

Experiment 8

Stretching Points

Name ______________________________

Partner(s) ______________________________

Mathematical Analysis

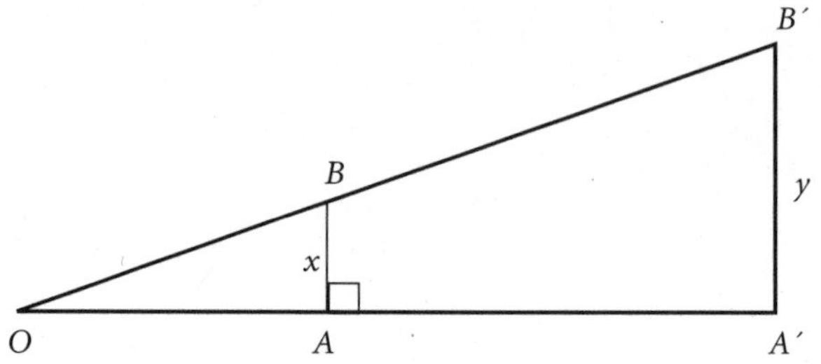

B'

B

y

x

O A A'

Name two similar triangles. ____________ ____________

Use this similarity to fill in the following blanks:

$$\frac{y}{x} = \frac{OA'}{\quad} = \frac{\quad}{OB}$$

All have the (constant) value K, the scale factor of the triangles.

Write an equation for the length.

$y =$ ______________________________

(in terms of x and K)

For your rubber band, what is the value of K? ____________

How did you determine this value?

____

____

____

For your rubber band, substitute your value of K and write y as a function of x.

Stretched length $= y =$ ____________

Graph your function (if you have a TI-82, use (Y=) and (GRAPH); if you have a TI-81, use DrawF). Do your data points lie *close* to the graph? If not, go back and determine the source of your error. Add the graph of the function to the Data Graph on the Collect the Data sheet.

Name ____

Partner(s) ____

Answer the following questions. Show your work. If you have not entered the function as **Y1**, do so now and graph it.

Interpret Your Findings

1. If the line segment from the knot to the base line is 10 units long, what is the length of the line segment from the end of the band to the base line? ________

   If the line segment from the knot to the base line is 100 units long, what is the length of the line segment from the end of the band to the base line? ________

2. If O is still on line L but is farther away from A, how does the function change?

   ____

   ____

   How did you decide on your answer?

   ____

   ____

3. Move O off line L. Now how does the function change?

   ____

   ____

4. If the knot had been exactly in the middle of the band, what would the equation of y be?

   ____

5. Mark where the knot would be if the resulting stretch equation were $y = 4x$.

   ____

   O

6. Mark where the knot would be if the resulting stretch equation were $y = (4/3)x$

   ____

   O

Experiment 9

Little House

Teaching Notes

In this experiment, students form the facade of a simple "house": a rectangle surmounted by an isosceles triangle. The base of the house is the *independent variable,* and the area of the facade is the *dependent variable.* The perimeters, R of the rectangle and P of the triangle, are fixed. This experiment is extremely challenging unless students have completed Experiments 3 and 4.

Key Concepts

Area of a triangle = $\frac{1}{2}$ base $\times$ height

Area of a rectangle = base $\times$ height

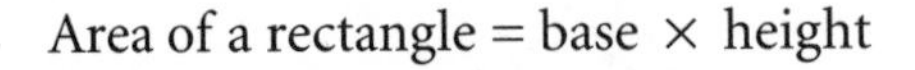

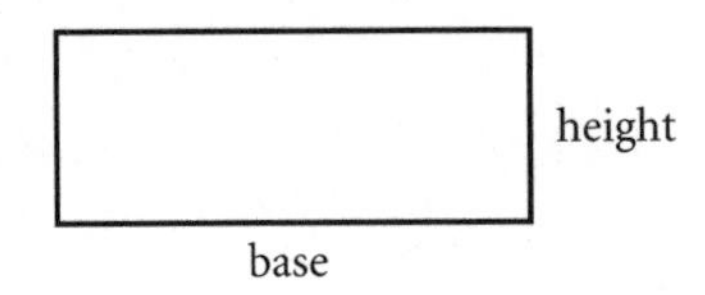

Perimeter of a triangle = the sum of the sides

Perimeter of a rectangle = 2 × (base + height)

Pythagorean Theorem: $a^2 + b^2 = c^2$

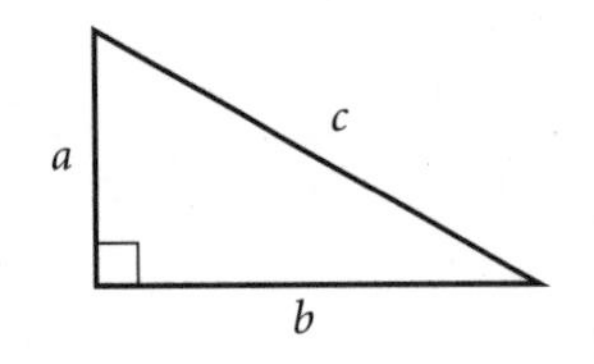

Equipment

graph paper (see page 131) with cardboard backing, 1 sheet per group

lengths of dark string, 14"–20", 2 per group

Tie the ends of each length to make a loop. The two loops for each group should be of different lengths and different colors.

pushpins (to anchor the corners), 5 per group

transparency of the graph paper

overhead projector

long pipe cleaners, 2 for teacher

Form pipe cleaners into loops—one into a rectangle and the other into an isosceles triangle, both with a base of about 3".

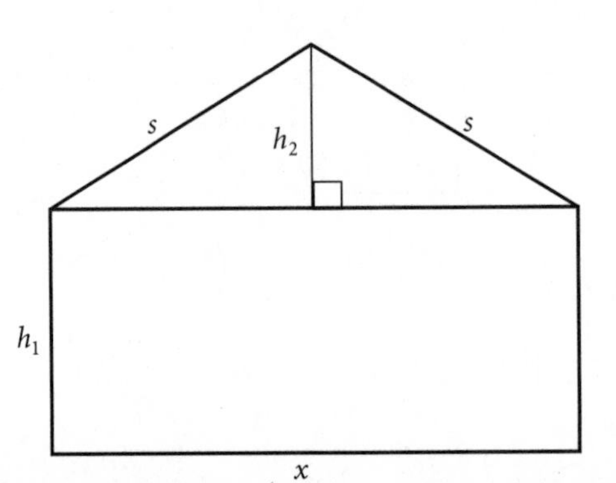

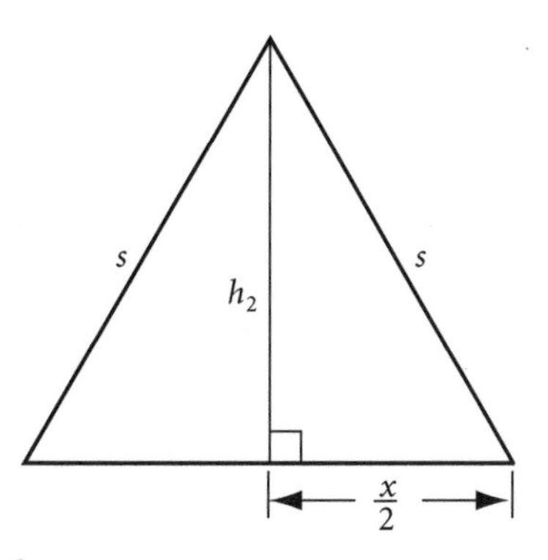

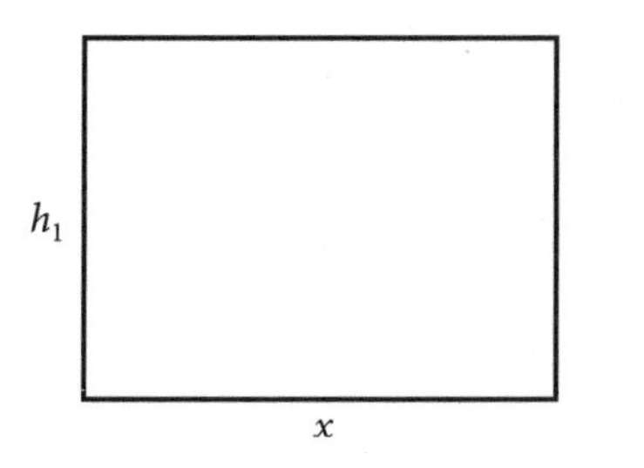

Procedure

Demonstrate at the overhead projector. Build a house from the pipe-cleaner rectangle and triangle on the graph-paper transparency. Units are the number of squares on the graph paper.

Have students use the graph paper as a guide to decide on the width of their house. After forming the rectangle, they build the roof by superimposing its base on the top of the rectangle, then pulling on the vertex. They measure the heights of both the rectangle and the triangle, then calculate the total area. Encourage them to use at least one short rectangle, one tall rectangle, and one "squarish" rectangle as the base of the house. Once they have measured six or more house fronts, groups plot the points (x, y), where y = total area on a graphing calculator. After plotting the points, students find the equation of the area function, then verify their analyses by graphing the function and observing how closely it fits the plotted points.

Mathematical Analysis

If the perimeter of the roof is P, then $P = 2s + x$ and $s = \dfrac{P - x}{2}$.

Using the Pythagorean Theorem: $\left(\dfrac{x}{2}\right)^2 + h_2^{\,2} = s^2$.

Solving for h_2 gives $h_2 = \sqrt{s^2 - \left(\dfrac{x}{2}\right)^2}$.

Substituting for s gives $h_2 = \sqrt{\left(\dfrac{P-x}{2}\right)^2 - \left(\dfrac{x}{2}\right)^2}$.

The area of the triangle is $y = \dfrac{1}{2}x\sqrt{\left(\dfrac{P-x}{2}\right)^2 - \left(\dfrac{x}{2}\right)^2} = \dfrac{x}{4}\sqrt{P^2 - 2Px}$.

If the perimeter, R, of the rectangle is given by $R = 2(x + h_1)$, then, solving for h_1: $h_1 = \dfrac{R - 2x}{2}$.

Substituting into the area formula, area = $x \cdot h_1$: area = $y = x\left(\dfrac{R-2x}{2}\right)$

Combining the two areas produces the following formula for the area of the facade: $y = \dfrac{x}{4}\sqrt{P^2 - 2Px} + \dfrac{x}{2}(R - 2x)$.

Extension

Ask students to conjecture the shape of the house of maximum area if the loops had been identical. Have them rewrite the area function with $R = P$, then choose a value for R and graph the function. They then make two loops of the same length and build the house of maximum area from them. Ask, "Is the base a square?" "Is the roof an equilateral triangle?"

Experiment 9

Little House

Name ______________________________

Partner(s) ______________________________

Collect the Data

Key Concepts

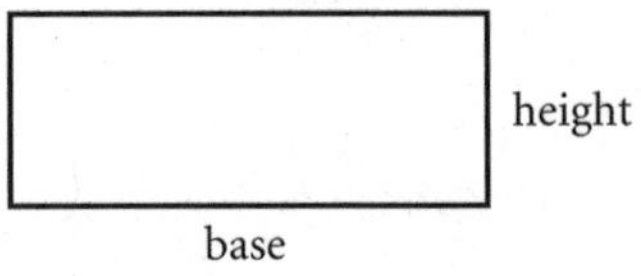

Area of a triangle = $\frac{1}{2}$ base × height

Area of a rectangle = base × height

Perimeter of a triangle = the sum of the sides

Perimeter of a rectangle = 2 × (base + height)

Pythagorean Theorem: $a^2 + b^2 = c^2$

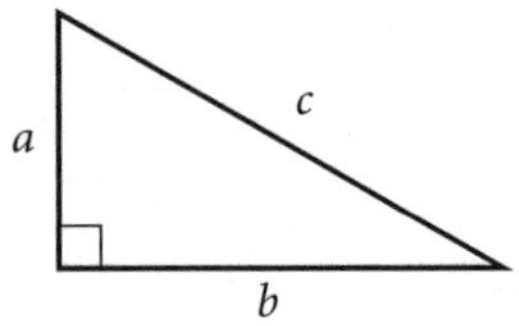

height

base

a

c

b

Describe the experiment.

____

____

____

____

____

Data Collection

x Base	Rectangle Height	Roof Height	Rectangle Area	Roof Area	Total Area

Points To Be Plotted

x	y

Enter the points as data points in your calculator, then plot them. Copy the points from the calculator display to the screen diagram below. Record the screen ranges, and label the axes.

Data Graph

Xmin = ____________

Xmax = ____________

Ymin = ____________

Ymax = ____________

Experiment 9

Little House

Name ______________________________

Partner(s) ______________________________

Mathematical Analysis

s s h_2 h_1 y x

Refer to the diagram. If the perimeter of the roof section is P, then

$P =$ ______________________________
(in terms of x and s)

Solve for s: $s =$ ______________________________
(in terms of x and h_2)

Use the Pythagorean Theorem to find s^2:
$s^2 =$ ______________________________
(in terms of x and h_2)

Solve for h_2: $h_2 =$ ______________________________
(in terms of x and s)

Now substitute for s: $h_2 =$ ______________________________
(in terms of x and P)

Now, express the area of the roof portion of the house:
Area $= y =$ ______________________________
(in terms of x and h_2)

Substitute for h_2: Area $= y =$ ______________________________
(in terms of x and P)

Use the same diagram. With R as the perimeter of the rectangle portion of the house, express the perimeter: $R =$ ______________________________
(in terms of x and h_1)

Solve for h_1: $h_1 =$ ______________________________
(in terms of x and R)

Express the area of the rectangle: $y =$ ______________________________
(in terms of x and h_1)

Combine the two areas into one function: Area $= y =$ ______________________________
(in terms of x, P, and R)

For your houses: $R =$ ____________ $P =$ ____________

Using these values, your house has the area function: Area $= y =$ ____________

Graph your function (if you have a TI-82, use (Y=) and (GRAPH); if you have a TI-81, use DrawF). Do your data points lie *close* to the graph? If not, go back and determine the source of your error. Add the graph of the function to the Data Graph on the Collect the Data sheet.

Name ____________________

Partner(s) ____________________

Answer the following questions. Show your work. If you have not entered the function as **Y1**, do so now and graph it.

Interpret Your Findings

1. Use (TRACE) to find the maximum possible value of the area of a house made with your loops.

   The maximum area is __________ . It occurs when $x =$ __________ .

2. Suppose you had switched the two loops, making the roof with a loop of length R and the rectangle with a loop of length P. What would your area function be now?

   ____

3. Clear the screen, and graph the function in question 2. Use (TRACE) to find the following:

   The maximum area is __________ .

   It occurs when $x =$ __________ .

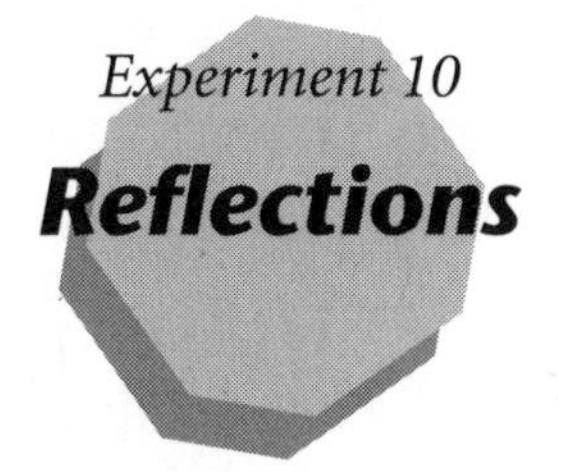

Experiment 10

Reflections

Teaching Notes

In this experiment, students open and close a pair of hinged mirrors to see the regular polygons that are formed by the reflection of a line. The number of sides of the polygon seen in the mirrors is the *independent variable,* and the angle between the mirrors is the *dependent variable.*

Key Concept

Sum of the exterior angles of a polygon = 360°

Equipment

small mirrors of the same magnification, 2 per group

The mirrors should be "hinged" with transparent tape. Acrylic "locker mirrors" can be cut easily by first scoring, then rapping the mirror sharply.

circular protractors, 1 per group

You may want to make the circular protractors by copying the blackline master on page 132 onto overhead-transparency sheets. Mark the center dot and the 0-degree line in color.

straightedges, 1 per group

blank paper, 1 sheet per group

sharp pencils, 1 per group

Procedure

In groups of two, students draw a 4- to 5-inch line across the middle of a piece of paper. They place the mirrors as shown, open to about 150°. Gradually, they close the mirrors until they see what appears to be an equilateral triangle formed by the base line and its reflections in the mirrors.

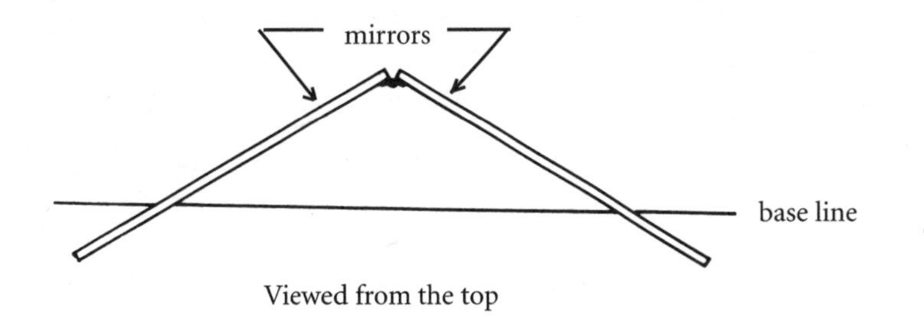

Viewed from the top

With a pencil, students trace along the inside edges of the mirrors. They remove the mirrors and, if necessary, extend the lines they have drawn until they meet. Using the protractor, students measure the angle formed by the mirrors. (If the mirrors are the same height and meet exactly, students can set the protractor on top of them and measure the angle directly.)

Students replace the mirrors and close them until a regular four-sided polygon appears, then again measure the angle of the mirrors. Have students collect at least six data points by adjusting the mirrors until they see a regular pentagon, hexagon, heptagon, and so on.

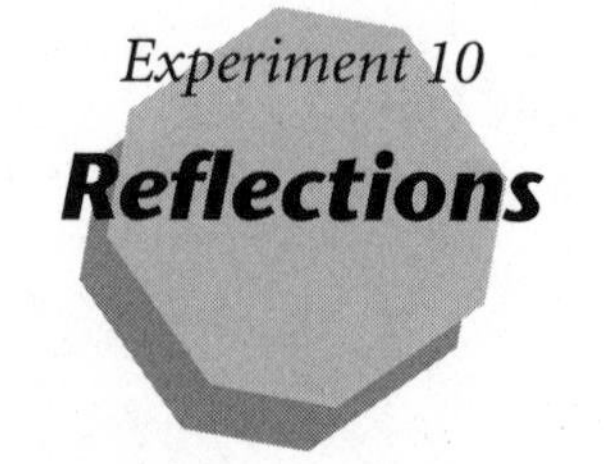

Teaching Notes, page 2

Mathematical Analysis

If the mirrors make an angle of y (measured in degrees), then each base angle, ß, of the triangle—an acute angle made by a mirror and the base line—is $\beta = \frac{180 - y}{2}$.

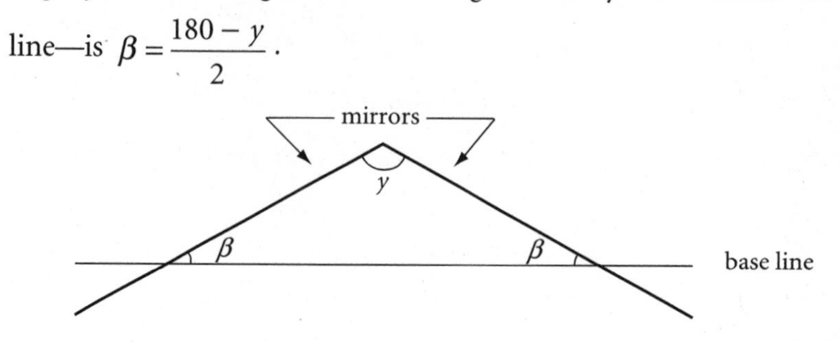

The mirror bisects the angle made by the base line and its reflections. Thus the exterior angle, θ, of the polygon = $180 - (180 - y) = y$.

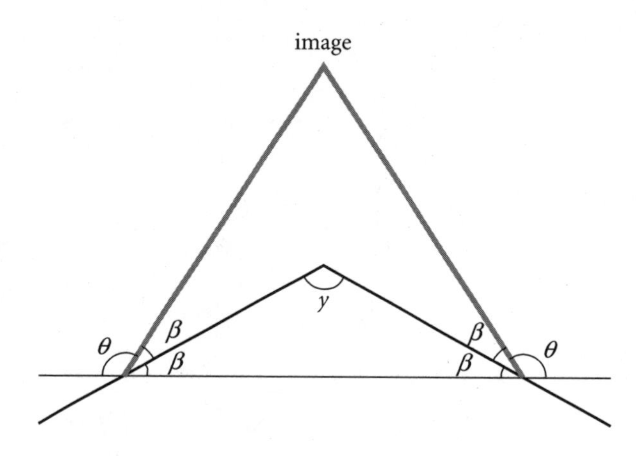

The sum of the exterior angles of a polygon is 360°. There are x exterior angles of y degrees: $xy = 360$, or $y = \frac{360}{x}$.

Experiment 10

Reflections

Name ______________________________

Partner(s) ______________________________

Key Concept

Sum of the exterior angles of a polygon = 360°

Collect the Data

Describe the experiment.

____

____

____

____

____

____

Data Collection

Sides	Angle

Points To Be Plotted

x	y

Enter the points as data points in your calculator, then plot them. Copy the points from the calculator display to the screen diagram below. Record the screen ranges, and label the axes.

Data Graph

Xmin = ____________

Xmax = ____________

Ymin = ____________

Ymax = ____________

Experiment 10

Reflections

Name ______________________________

Partner(s) ______________________________

Mathematical Analysis

mirrors

y

β β

base line

image

y

θ α θ

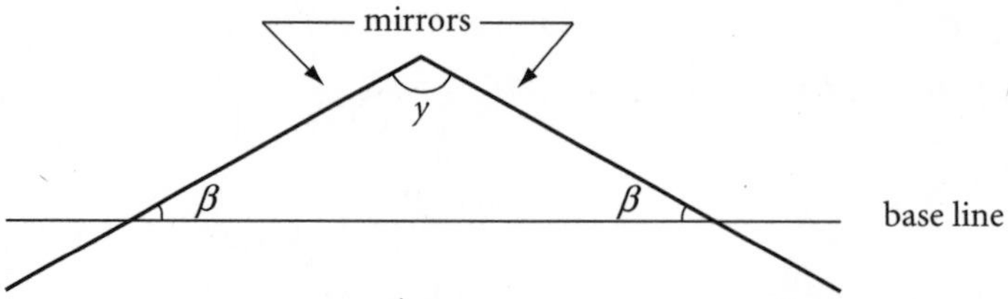

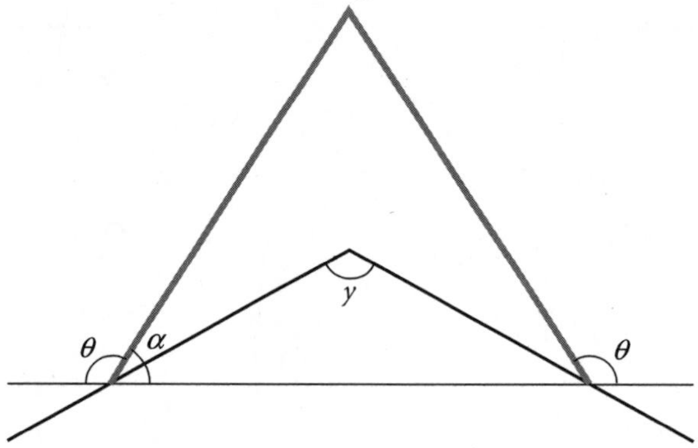

Assume the polygon has x sides.

If the mirror angle is y (measured in degrees), each of the base angles made by the mirror and the base line = ______________________________
(in terms of y)

The mirror bisects the angle, α, made by the base line and its image.

The angle α = ______________________________
(in terms of y)

The exterior angle, θ, is made by the image and the base line.

The angle θ = ______________________________
(in terms of y)

The sum of the exterior angles is $x\theta$ = ______________ = ____________°
(in terms of y)

Now, solve for y: y = ______________________________
(in terms of x)

Graph your function (if you have a TI-82, use (Y=) and (GRAPH); if you have a TI-81, use DrawF). Do your data points lie *close* to the graph? If not, go back and determine the source of your error. Add the graph of the function to the Data Graph on the Collect the Data sheet.

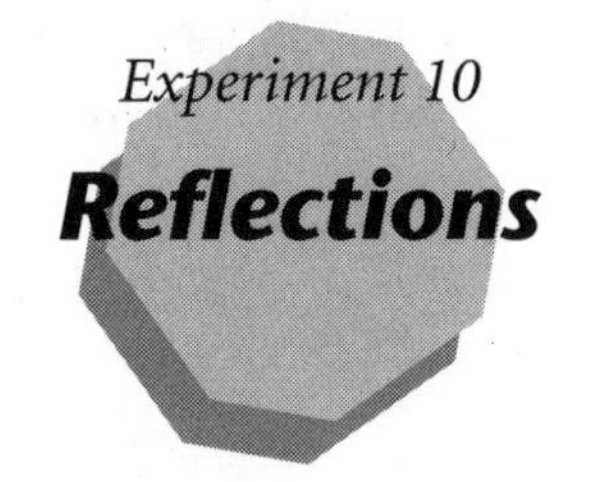

Experiment 10

Reflections

Name ______________________________

Partner(s) ______________________________

Interpret Your Findings

Answer the following questions. Show your work. If you have not entered the function as **Y1**, do so now and graph it.

1. Use (TRACE) and (ZOOM) or algebra to find the number of sides you would see if the angle between the mirrors were 24°. ______________

2. What is the value of your function for $x = 2$? _______ What is the physical meaning of this value?

   ____
   ____
   ____
   ____
   ____

3. What kind of polygon would you expect to see if the angle were 80°?

   ____

   Now, make an 80° angle with the mirrors. What *do* you see?

   ____

   Try to draw a diagram for this situation.

Experiment 11

Looking Down

Teaching Notes

In this experiment, a small mirror is hung below eye level, and a yardstick is placed on the floor so that it extends from the wall. Students stand at distance x from the wall, look into the mirror, and observe the reflection of the mark on the yardstick that appears at the center of the mirror. The distance of the viewer from the wall is the *independent variable,* and the ruler mark seen in the mirror is the *dependent variable.*

Key Concept

Similar triangles: Corresponding sides are proportional

Equipment

small mirrors, 1 per group

tape (or other way to affix mirror to wall)

markers, 1 per group

yardsticks, 1 per group

Procedure

Have students, working in groups of two or three, place the mirror on the wall 18 to 30 inches from the ground. With a marker, they draw a horizontal line across the middle of the mirror. They extend the yardstick out from the wall along the floor, with the zero point against the wall. When they stand sufficiently close to the mirror, they will see a portion of the yardstick reflected in the mirror. The point on the yardstick that is reflected at the line on the mirror is the point y.

Mathematical Analysis

The angle of reflection (θ) is equal to the angle of incidence. If the observer's height is p and the center of the mirror is a distance m from the floor, then, by similar triangles,

$$\frac{y}{x} = \frac{m}{p-m}.$$

In a more common form,

$$y = \frac{m}{p-m}x.$$

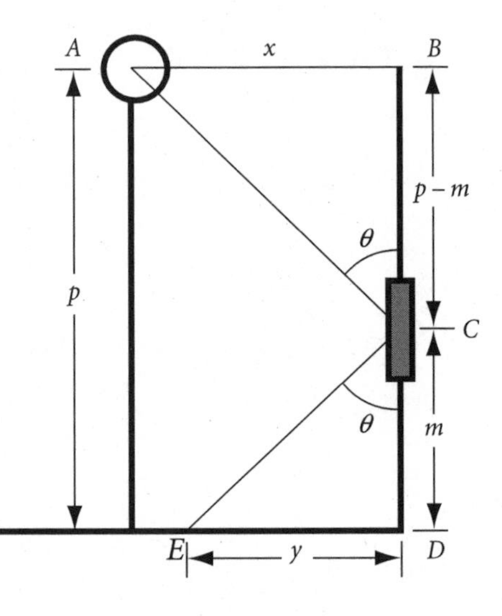

Experiment 11

Looking Down

Name ______________________________

Partner(s) ______________________________

Key Concept

Similar triangles: Corresponding sides are proportional

Collect the Data

Describe the experiment.

____

____

____

____

____

____

Units used: __________

Data Collection

Distance	Mark

Points To Be Plotted

x	y

Enter the points as data points in your calculator, then plot them. Copy the points from the calculator display to the screen diagram below. Record the screen ranges, and label the axes.

Data Graph

Xmin = __________

Xmax = __________

Ymin = __________

Ymax = __________

Experiment 11

Looking Down

Name ______________________________

Partner(s) ______________________________

Mathematical Analysis

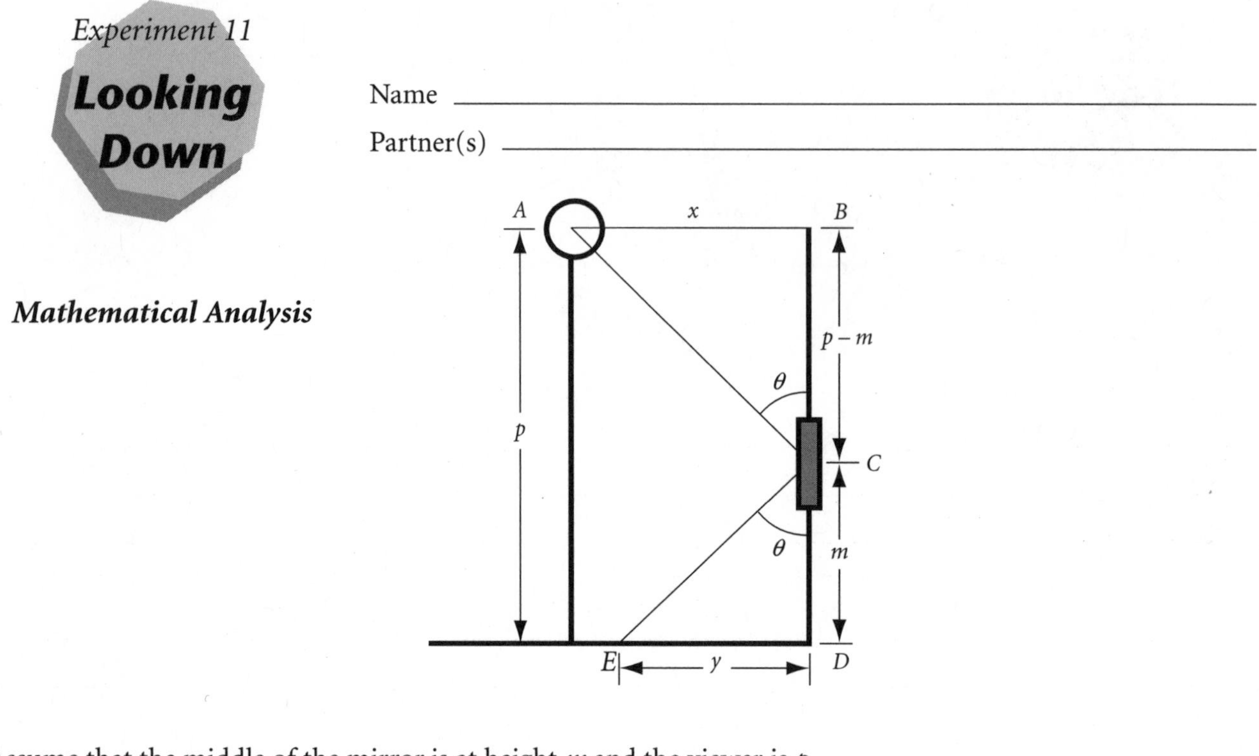

Assume that the middle of the mirror is at height m and the viewer is p units tall.

What triangle is similar to ΔCDE? ______________________________

Complete the following equation: $\dfrac{y}{x}$ = ______________________________

Solve for y, which is the mark seen in the mirror: y = ____________________
(in terms of x, p, and m)

In your experiment, m = ____________________ and p = ____________________

Substitute your values of m and p to obtain the function for your setup:

y = ______________________________
(in terms of x)

Graph your function (if you have a TI-82, use (Y=) and (GRAPH); if you have a TI-81, use DrawF). Do your data points lie *close* to the graph? If not, go back and determine the source of your error. Add the graph of the function to the Data Graph on the Collect the Data sheet.

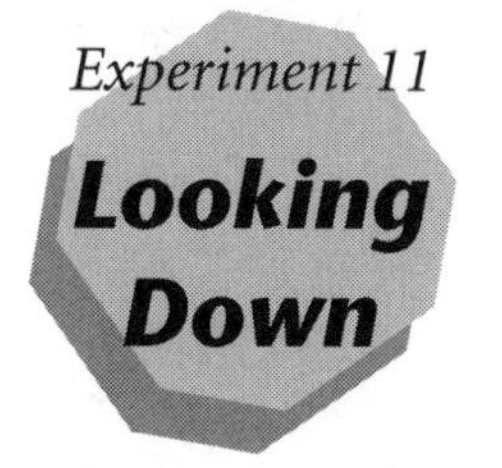

Name ______________________________

Partner(s) ______________________________

Answer the following questions. Show your work. If you have not entered the function as **Y1**, do so now and graph it.

Interpret Your Findings

1. Use (TRACE) or algebra to find the value of x for which $y = m$. ________

2. If the mirror had been 4 units higher, would the graph have been steeper or flatter? How do you know?

____

____

____

3. Did your answer to question 2 depend on the height of the person doing the viewing? Why or why not?

____

____

____

4. Another group came up with the equation $y = \frac{1}{2}x$. The viewer was exactly 63 inches tall. Find the height of the center of the mirror.

____

5. Which variable, p or m, would have the most effect on the slope of the line? ________ Explain your answer.

____

____

____

6. What would your equation be if you had recorded your measurements in centimeters instead of inches (or inches if you used centimeters)?

____

Teaching Notes

In this experiment, students investigate the sum of the interior angles of a polygon. The *independent variable* is the number of sides, and the *dependent variable* is the sum of the interior angles.

Key Concept

Interior and exterior angles of a polygon are supplementary

Equipment

pencils, 1 per student

straightedges, 1 per student

paper, several sheets per student

transparent circular protractors, 1 per student

You may want to make the circular protractors by copying the blackline master on page 132 onto overhead-transparency sheets. Mark the center dot and the 0-degree line in color.

Procedure

Groups of three students work well, as the group can either average the three measurements or use the middle one. Starting with $x = 4$, each student in the group draws a polygon of x sides using a straightedge and making clearly defined corners. Encourage students to use creativity in drawing their polygons. Students should mark the interior angles, measure them, and find their sum.

The data point to be plotted is (x, A), where A, the angle sum, is the average of the sums—or the middle sum—found by the group. Students should collect data for $x = 4, 5, 6, 7$, and one or two larger values on which the group agrees. One of the polygons, for example, could be the school's initial. The polygon should not have "holes":

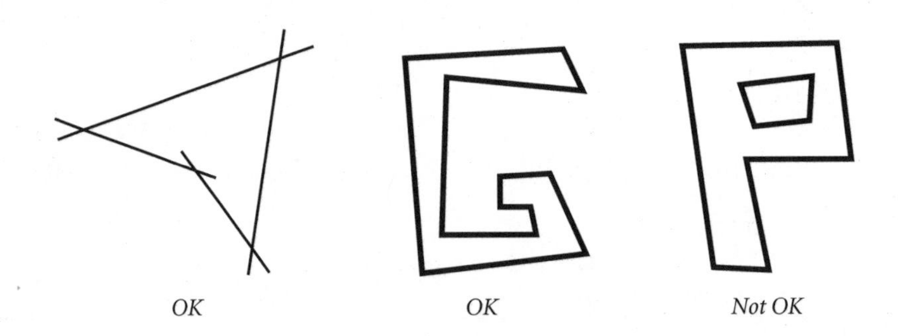

Mathematical Analysis

The sum of the interior angles is found by thinking about the total angle you would turn through if you had "walked" around the polygon. Suppose the polygon has x sides; it also has x vertices. If you hiked completely around the polygon and turned to face your starting direction, you would make one complete turn of 360°. Additionally, at each vertex, the sum of the turning angle and interior angle is 180°.

Experiment 12

Polygons

Teaching Notes, page 2

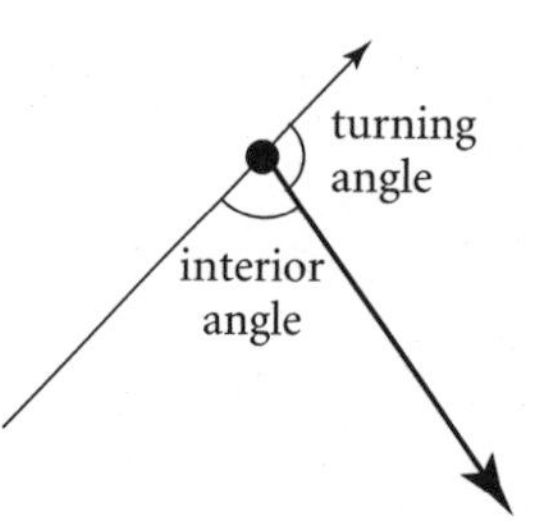

The sum of all (turning angle + interior angle) sums is $180x$.

(sum of turning angles) + (sum of interior angles) = $180x$

360 + (sum of interior angles) = $180x$

(sum of the interior angles) = $180x - 360$

Experiment 12

Polygons

Name ______________________________

Partner(s) ______________________________

Key Concept

Interior and exterior angles of a polygon are supplementary

Collect the Data

Describe the experiment.

____

____

____

____

____

Data Collection

Number of Sides	Sum of Angles

Points To Be Plotted

x Number of Sides	y Sum of Angles

Enter the points as data points in your calculator, then plot them. Copy the points from the calculator display to the screen diagram below. Record the screen ranges, and label the axes.

Data Graph

Xmin = __________

Xmax = __________

Ymin = __________

Ymax = __________

Name ______________________________

Partner(s) ______________________________

Mathematical Analysis

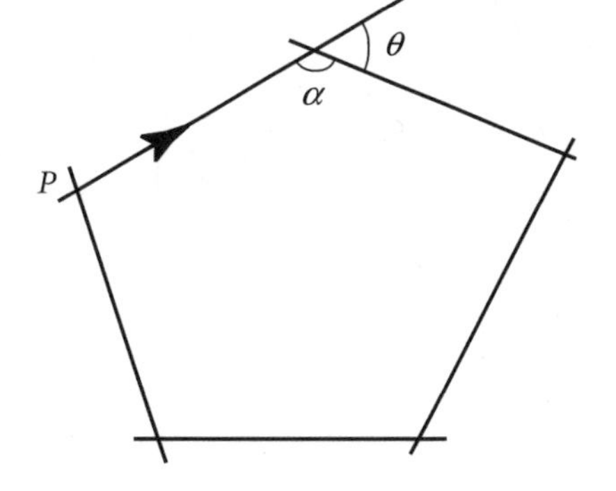

Suppose you were to start at P, headed in the direction of the arrow. Walk completely around the polygon, stopping when you return to P and are *headed in the starting direction.*

a. At every corner, you have turned through the angle corresponding to θ. What is the total angle through which you have turned?

   Sum of all θ = ______________________________

b. At each corner, what is the sum $(\theta + \alpha)$? ____________________

   What is the total of all of the $(\theta + \alpha)$ sums?
   Sum of all $(\theta + \alpha)$ = ______________________________

c. Use the fact that the sum of all $(\theta + \alpha)$ = sum of all θ + sum of all α and substitute your results from part a and part b to find the sum of all α for this picture.

   Sum of all α = ______________________________

d. In the picture above, how many sides are there? ____________________

   What is the sum of the interior angles? ____________________

   If the number of sides were x, what would the sum of the angles be?

   Sum of the angles = y = ______________________________
   (in terms of x)

Graph your function (if you have a TI-82, use (Y=) and (GRAPH); if you have a TI-81, use DrawF). Do your data points lie *close* to the graph? If not, go back and determine the source of your error. Add the graph of the function to the Data Graph on the Collect the Data sheet.

Interpret Your Findings

Name ______________________________

Partner(s) ______________________________

Answer the following questions. Show your work. If you have not entered the function as **Y1**, do so now and graph it.

1. Draw a 10-sided polygon with all right angles. Mark the interior angles.

   Every interior angle is either 90° or ____________°.

   What is the sum of the angles in your figure? ____________

2. Draw an 11-sided polygon with as many right angles as possible. Can you make all the angles either right angles or 45° angles? ____________

   What is the sum of the interior angles? ____________

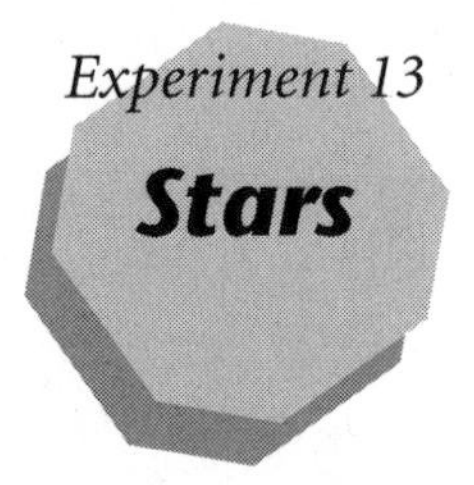

Experiment 13

Stars

Teaching Notes

This experiment builds on the results of the previous experiment, Polygons. Students investigate the sum of the interior angles of a polygon and the sum of the "vertex" angles of a *star polygon* (a polygon whose sides cross themselves). Groups of three work well, as the group can either average the three measurements or use the middle one. The *independent variable* is the number of sides, and the *dependent variable* is the sum of the angles at the vertices.

Key Concept

Interior and exterior angles of a polygon are supplementary

Equipment

pencils, 1 per student

straightedges, 1 per student

Stars Sheet (see page 88), 1 per student

transparent circular protractors, 1 per student

You may want to make the circular protractors by copying the blackline master on page 132 onto overhead-transparency sheets. Mark the center dot and the 0-degree line in color.

Procedure

Working in groups of three, students make star polygons by putting n points on a circle on the Stars Sheet, connecting every third point, and measuring the sum of the angles at the points of the star. The points need not be evenly spaced. If the points have not all been "hit" on the pencil's path, the students should lift the pencil and start again at the next unused point. Students make stars with 7, 8, 10, and 11 vertices.

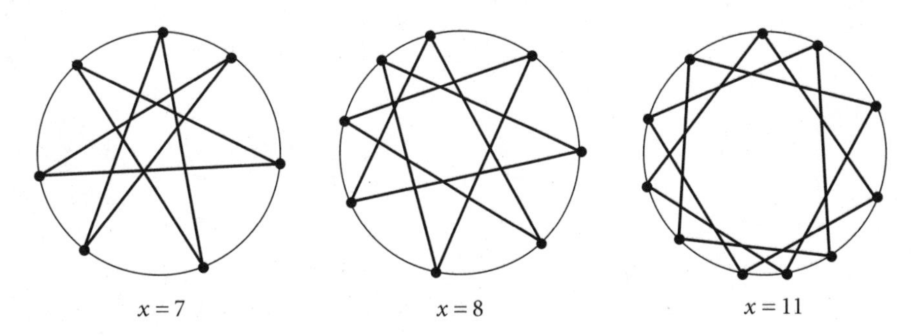

$x = 7$ $x = 8$ $x = 11$

Mathematical Analysis

The sum of the interior angles is found by thinking about the total angle turned through as if you "walked" around the star. Suppose the star has x points, with every third point connected. If 3 *is not* a divisor of the total number of points on the circle, the star can be drawn using all the points without lifting the pencil. If you hiked completely around the star and turned to face your starting direction, you would have made three full turns (3×360). At each of the vertices, the sum of the turning angle and interior angle is 180°.

Teaching Notes, page 2

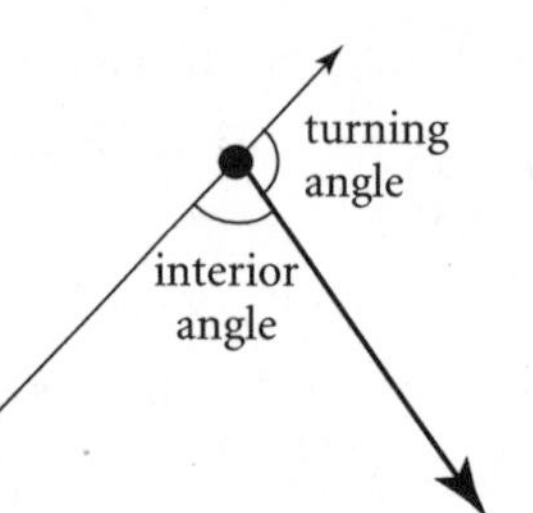

The sum of all (turning angle + interior angle) sums is $180x$

(sum of turning angles) + (sum of interior angles) = $180x$

$3 \cdot 360$ + (sum of interior angles) = $180x$

or,

(sum of the interior angles) = $180x - 3 \cdot 360$

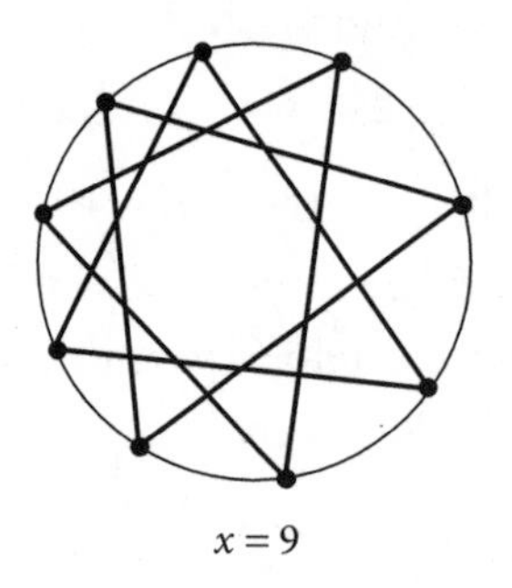

$x = 9$

For $x = 11$, $3 \cdot 360$ + (sum of angles at star points) = $11 \cdot 180$. The sum of the angles is 900°.

If the pencil must be lifted from the paper, 3 *is* a factor of *x*, and three separate polygons are drawn, each with $\frac{x}{3}$ sides. From the Mathematical Analysis for Experiment 12, the sum of the interior angles of each polygon is $180\left(\frac{x}{3} - 2\right)$. Taking all three polygons together gives the "nonconnected" star with (sum of interior angles) = $3 \cdot 180\left(\frac{x}{3} - 2\right) = 180(x - 6)$.

Experiment 13

Stars

Name ______________________________

Partner(s) ______________________________

Key Concept

Interior and exterior angles of a polygon are supplementary

Collect the Data

Describe the experiment.

____

____

____

____

____

____

Data Collection

Number of Points	Sum of Angles

Points To Be Plotted

x Number of Points	y Sum of Angles

Enter the points as data points in your calculator, then plot them. Copy the points from the calculator display to the screen diagram below. Record the screen ranges, and label the axes.

Data Graph

Xmin = __________

Xmax = __________

Ymin = __________

Ymax = __________

Experiment 13

Stars

Mathematical Analysis

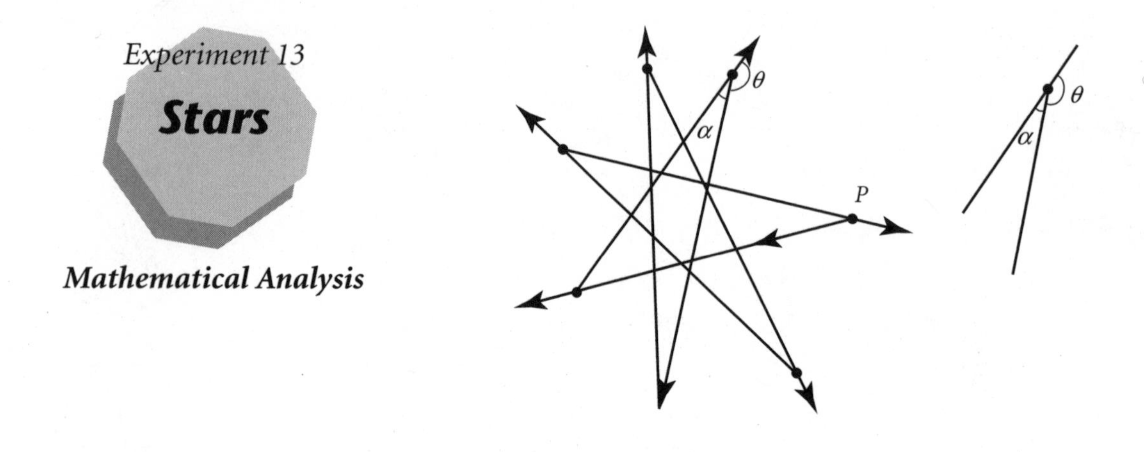

Suppose you were to start at *P*, headed in the direction of the arrow. Walk completely around the star, stopping when you return to *P* and are *headed in the starting direction.*

a. At every corner, you have turned through the angle corresponding to θ. What is the total angle through which you have turned?

Sum of all θ = ______________

b. At each corner, what is the sum $(\theta + \alpha)$? ______________

What is the total of all of the $(\theta + \alpha)$ sums?

Sum of all $(\theta + \alpha)$ = ______________

c. Use the fact that the sum of all $(\theta + \alpha)$ = sum of all θ + sum of all α and substitute your results from part a and part b to find the sum of all α for this picture.

Sum of all α = ______________

d. In this picture, how many sides are there? (*Side* refers to the line segment joining two vertices.) ______________

What is the sum of the interior angles? ______________

If the number of sides were *x*, what would the sum of the angles be?

Sum of the angles = y = ______________
(in terms of *x*)

Graph your function (if you have a TI-82, use [Y=] and [GRAPH]; if you have a TI-81, use DrawF). Do your data points lie *close* to the graph? If not, go back and determine the source of your error. Add the graph of the function to the Data Graph on the Collect the Data sheet.

Experiment 13

Stars

Name ______________________________

Partner(s) ______________________________

Interpret Your Findings

Answer the following questions. Show your work. If you have not entered the function as **Y1**, do so now and graph it.

1. Suppose, with x points, you had connected every point to the *first* point to its right. What kind of figure is this? ______________
   What is the sum of the vertex angles? ______________

2. On one of the empty circles on the Stars Sheet, mark seven points and connect every *second* point. On the other empty circle, mark eight points and connect every point to the *second* point to its right. For each figure, find the sum of the angles at the vertices.
   7 points ______________
   8 points ______________

3. Look at these functions for the sums of the vertex angles:
   $y = 180x - 1 \cdot 360$ (when every first point is connected)
   $y = 180x - 3 \cdot 360$ (when every third point is connected)

   What is a reasonable prediction for $y =$ angle sum if each point connects to the second point to its right (or to its left)?

   ____

   What does your formula predict for $x = 7$ when every second point is connected?

   ____

   What does your formula predict for $x = 8$ when every second point is connected?

   ____

4. Make a prediction for a star with 8 points, with every third point connected. Do a mathematical analysis for this star.

   ____

Stars Sheet

Draw six stars. Start with seven (or more) points and connect every third point. Save two circles for the Interpret Your Findings sheet.

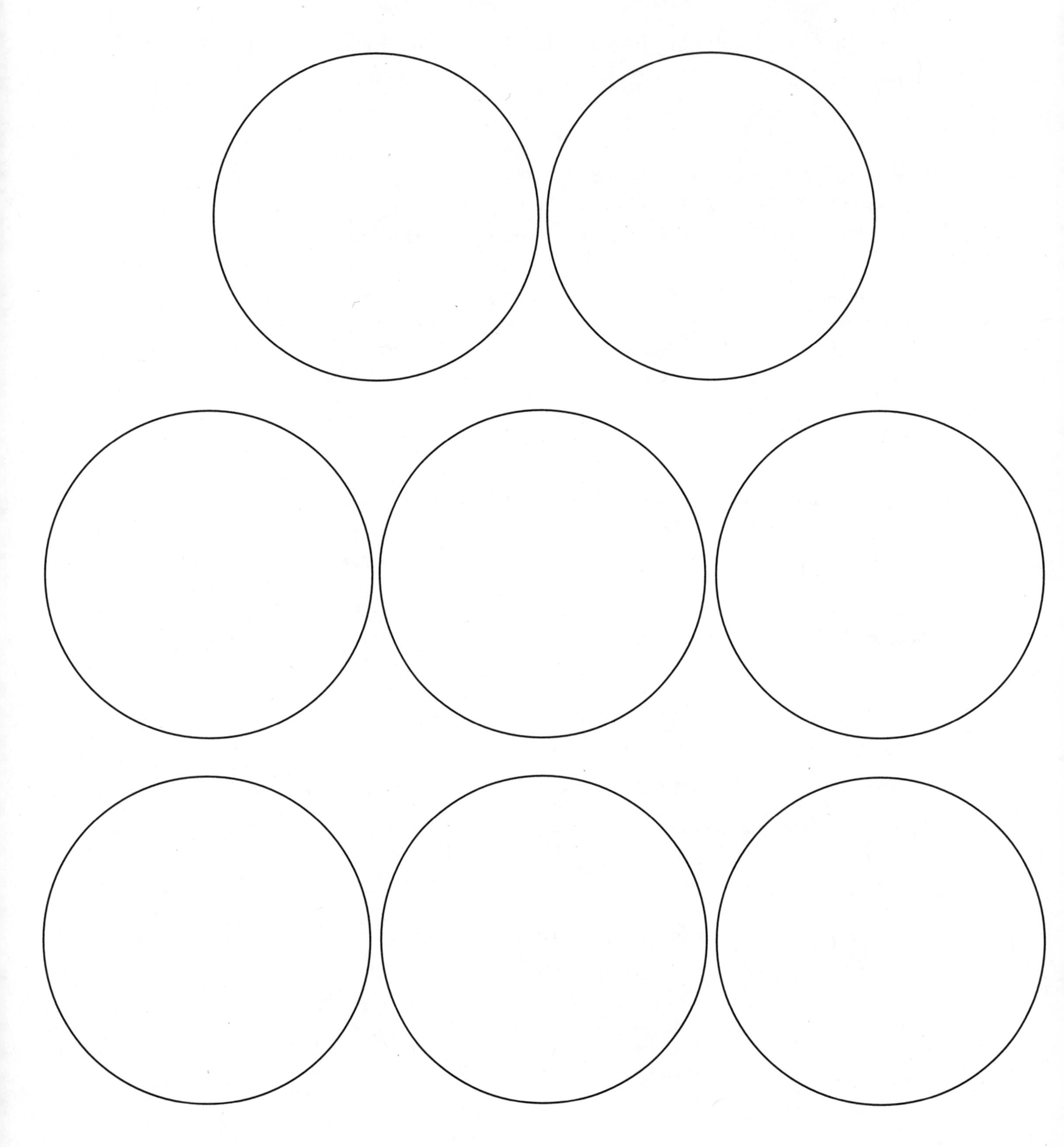

Experiment 14

Stretchy Isosceles Triangle

Teaching Notes

In this experiment, students measure three distances between knots (or ornaments) on a "growing" triangle. The *independent variable* is the height of the isosceles triangle, and the *dependent variable* is the distance between the knots.

Key Concepts

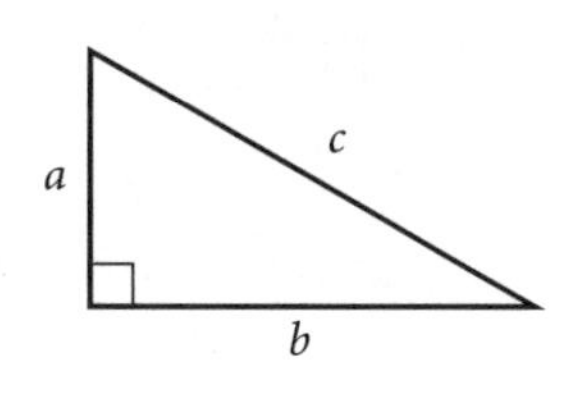

Similar triangles: Corresponding sides are proportional

Pythagorean Theorem: $a^2 + b^2 = c^2$

Equipment

identical rubber bands, 6 per group

pushpins, 2 per group

graph paper (see page 131) with cardboard backing, 1 sheet per group

rulers, 1 per group

Procedure

Students form two chains of three rubber bands each by looping the bands together at points *A*, *B*, *C*, and *D*. Working on the graph paper, they anchor one end of each chain along a base line. They loop the two loose ends over a pencil and stretch the bands to form an isosceles triangle of height *x*.

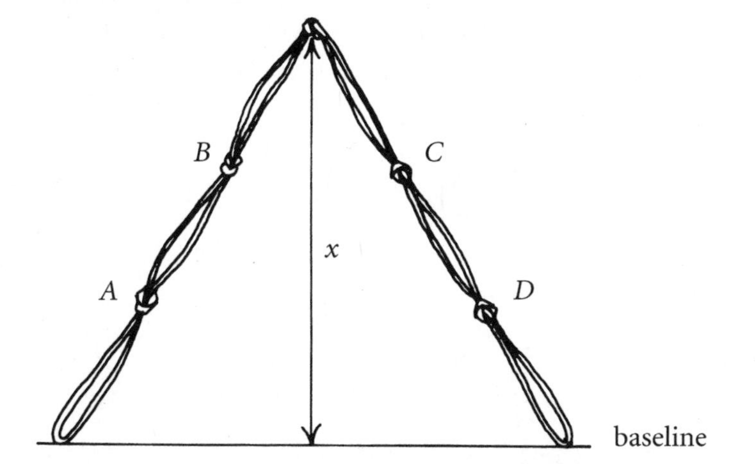

For different values of *x* (keeping the bands taut), students measure the distances *AD*, *BC*, and *AB* and record their measurements on the Collect the Data sheet.

Mathematical Analysis

The functions $y_1 = AD$ and $y_2 = BC$ are constant functions. If the fixed base of all the triangles is *b*, then, by applying similar triangles:

$$y_1 = \frac{2}{3}b \text{ and } y_2 = \frac{1}{3}b$$

The third function, $y_3 = AB$, is found by applying the Pythagorean Theorem:

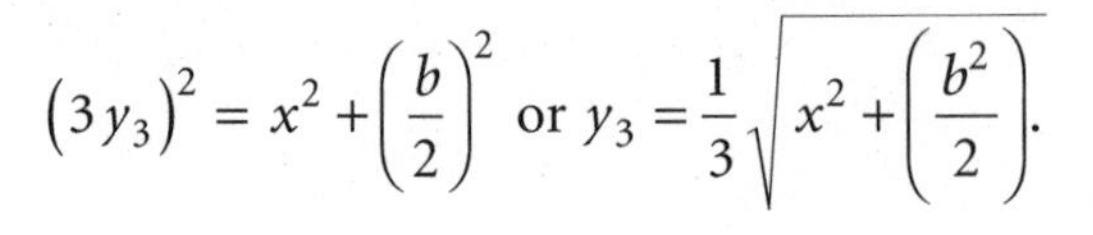

$$(3y_3)^2 = x^2 + \left(\frac{b}{2}\right)^2 \text{ or } y_3 = \frac{1}{3}\sqrt{x^2 + \left(\frac{b^2}{2}\right)}.$$

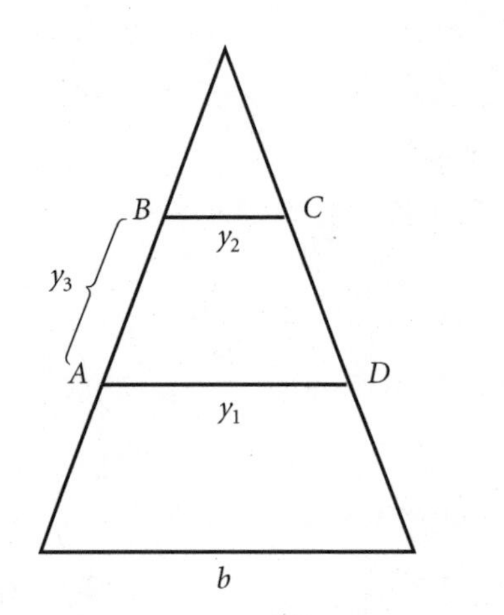

As they complete the Interpret Your Findings sheet, students should find:

Question 2. $AB = BC$ when the triangle is equilateral

Question 4. $AB = AD$ when each leg of the triangle is twice the base

It might be useful to demonstrate results 2 and 4 algebraically as a whole-class follow-up. To prove the result found in question 4, go through the following explanation:

$AB = AD$

$$\frac{1}{3}\sqrt{x^2 + \left(\frac{b}{2}\right)^2} = \frac{2}{3}b$$

$$x^2 + \left(\frac{b}{2}\right)^2 = 4b^2 \text{ or } x^2 = \frac{15}{4}b^2$$

For this value of x,

$$y_3 = \frac{1}{3}\sqrt{\frac{15}{4}b^2 + \frac{b^2}{4}} = \frac{2}{3}b$$

Hence, the length of side $= 3y_3 = 2b$.

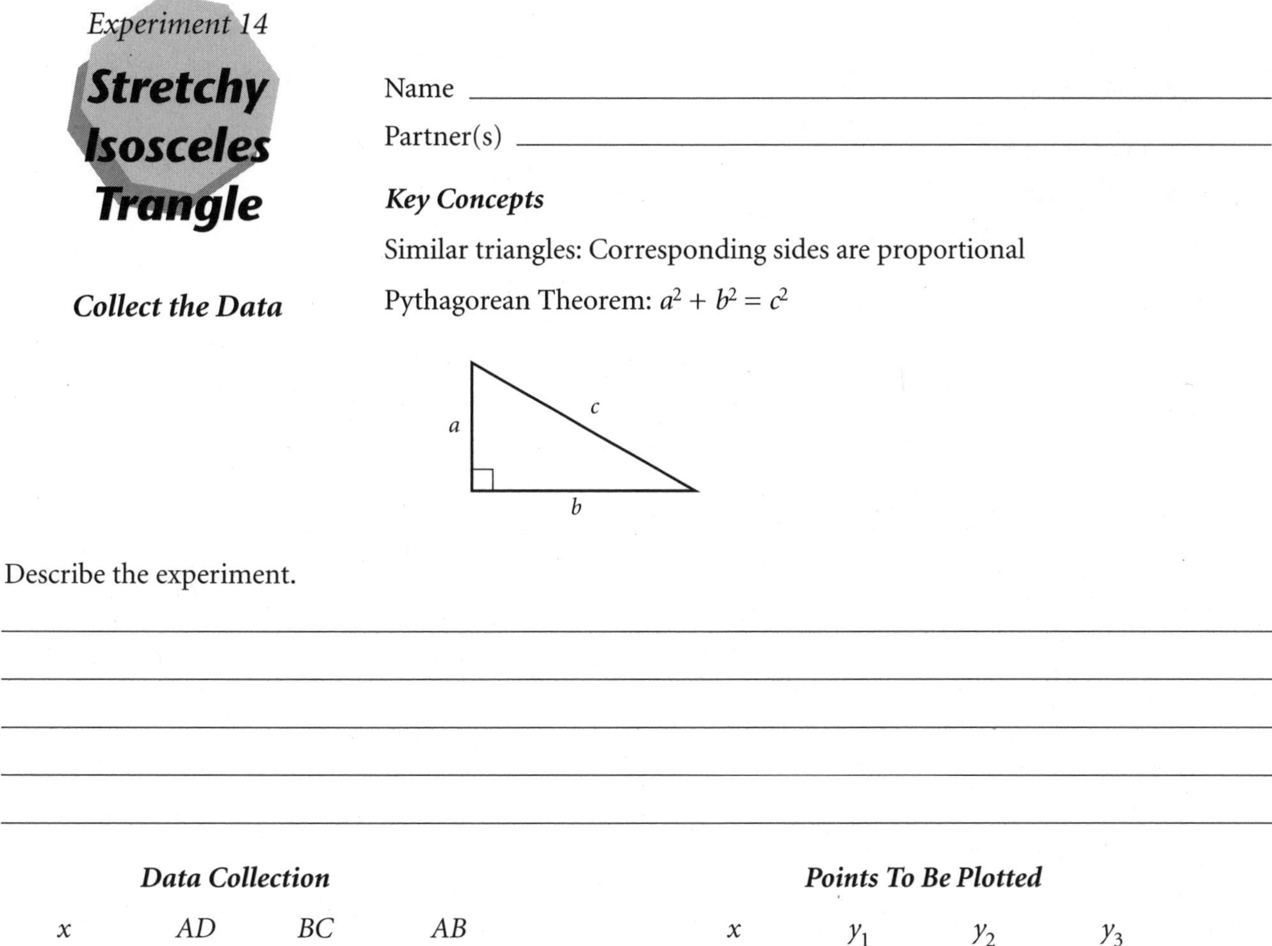

Experiment 14

Stretchy Isosceles Trangle

Name ______________________________

Partner(s) ______________________________

Key Concepts

Similar triangles: Corresponding sides are proportional

Pythagorean Theorem: $a^2 + b^2 = c^2$

Collect the Data

Describe the experiment.

____

____

____

____

____

Data Collection

x	AD	BC	AB

Points To Be Plotted

x	y_1	y_2	y_3

Enter the points (x, y_3) as data points in your calculator, then plot them. Copy the points from the calculator display to the screen diagram below. Record the screen ranges, and label the axes.

Data Graph

Xmin = ____________

Xmax = ____________

Ymin = ____________

Ymax = ____________

Experiment 14

Stretchy Isosceles Triangle

Name ______________________________

Partner(s) ______________________________

Mathematical Analysis

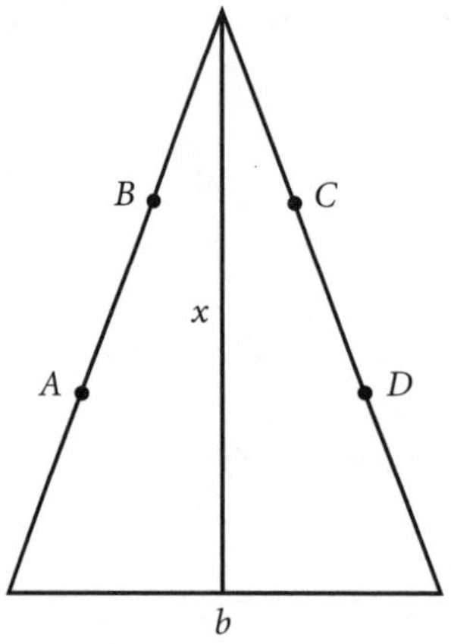

Assume the isosceles triangle has base b and height x. A and B divide one side of the triangle into three equal segments. C and D do the same for the other side. Use similar triangles to find expressions for the following:

$y_1 = AD =$ ______________ (in terms of b and x) $y_2 = BC =$ ______________ (in terms of b and x)

What is the value of b for all your triangles? ______________

For your triangles:

$y_1 = AD =$ ______________ $y_2 = BC =$ ______________

Now find an equation giving the distance AB in terms of x and b. Show your reasoning.

In terms of x and b, $y_3 = AB =$ ______________

Substitute your value of b: $y_3 =$ ______________

Graph your functions (if you have a TI-82, use (Y=) and (GRAPH); if you have a TI-81, use DrawF). Do your data points lie *close* to the graph? If not, go back and determine the source of your error. Add the graph of the function to the Data Graph on the Collect the Data sheet.

Experiment 14

Stretchy Isosceles Trangle

Name ____

Partner(s) _____________________________________

Answer the following questions. Show your work. If you have not entered the functions as **Y1**, **Y2**, and **Y3**, do so now and graph them.

Interpret Your Findings

1. Use (TRACE) or algebra to find the value of x for which $y_2 = y_3$. ________

2. Set up your experiment using the value of x you found in question 1. Is there an "interesting" triangle or angle involved? Make a scale drawing.

3. Find the value of x for which $y_3 = y_1$. Explain in detail the method you used.

   ____

   ____

   ____

4. Set up your experiment using the value of x you found in question 3. Make a scale drawing. What is the apparent relationship between the base and the length of a side?

   Verify your findings algebraically.

Experiment 15

School Flower Beds

Teaching Notes

In this experiment, students form a rectangle to enclose either a capital *M* or a capital *S*. The total of the lengths of the segments froming the letter is a constant. For the *S*, the width is the *independent variable.* For the *M*, the height is the *independent variable.* For both letters, the area of the enclosing rectangle is the *dependent variable.*

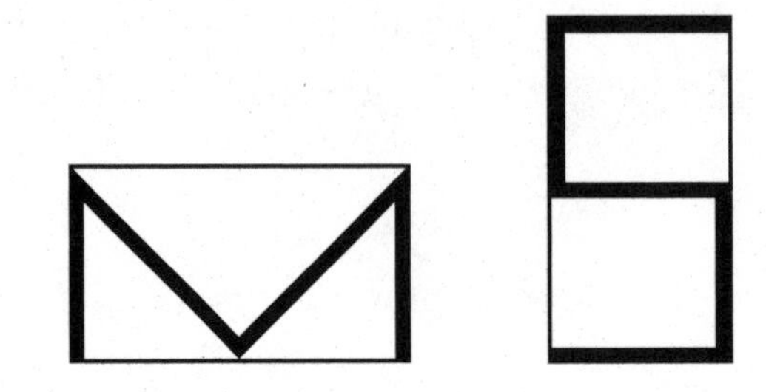

Key Concepts

Area of a rectangle = base × height

Pythagorean Theorem: $a^2 + b^2 = c^2$

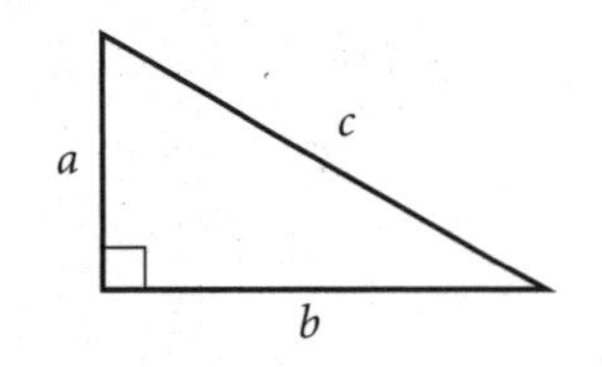

Equipment

string cord, 1 length per group

graph paper (see page 131) with cardboard backing, 1 sheet per group

circular protractors, 1 per group

You may want to make the circular protractors by copying the blackline master on page 132 onto overhead-transparency sheets. Mark the center dot and the 0-degree line in color.

pushpins, 6 per group

Procedure

The Metropolitan School District has decided to have student athletes plant flower beds in front of the school; the money saved on gardening will go toward funding the sports program. Each school team has been given the *same length* strip of red petunia seeds. The strip is narrow and can be thought of as a string of seeds (that is, as a line). With their seed strip, teams form one of the school's initials—*M* or *S*—then enclose it exactly in a rectangular plot of white petunias. The initial must be made in block form. To increase student enthusiasm, the school district has offered caps to the team whose rectangle has the largest total area.

After explaining the contest, ask whether students think the winning team will have planted an *M* or an *S*. You might decide to hold a mini-contest in the classroom. Let those who believe the *M* will be best work on an *M*; pair them with a group who choose the *S* and give the two groups strings of the same length.

Alternatively, introduce the *S* as a whole-class activity. Construct an *S* on the bulletin board, anchoring all corner points with pushpins. Ask for suggestions about the independent variable—the natural choice for x is the width of the *S*. Measure the width (x) and the height, and calculate the area (y). Generate three more data points by having different pairs of students construct an *S* at the bulletin board. Ask one pair for a "tall" *S* and one pair for a "short" *S*. Plot the points; their nonlinearity will be evident.

After the introduction, have students do the experiment using their length of string on the graph paper. For students working on the *M*, discuss what would be the best choice for the independent variable: "When you start planning the *M*, which dimension do you decide on first?" Since the height of the *M* is one of the dimensions of the flower bed, it is a logical choice for the independent variable.

Mathematical Analysis

For the *S*, the seed strip (the length of string) has a length L. The base of the rectangle is the independent variable x. $L - 3x$ is what remains of the seed strip for forming the two vertical parts of the letter. Thus the area of the enclosing rectangle is $x(L - 3x)$.

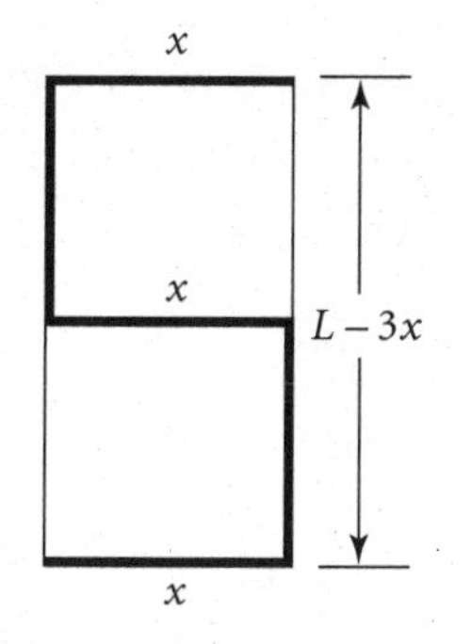

For the *M*, the seed strip again has a length L. The height of the rectangle is the independent variable x. Use half of the *M*, /|, to find the remaining dimensions. $\frac{L}{2}$ makes up this portion. Subtract x to find the length of the hypotenuse, and use the Pythagorean Theorem to solve for the other leg of the triangle:

$$\sqrt{\left(\frac{L}{2} - x\right)^2 - x^2}$$

Teaching Notes, page 3

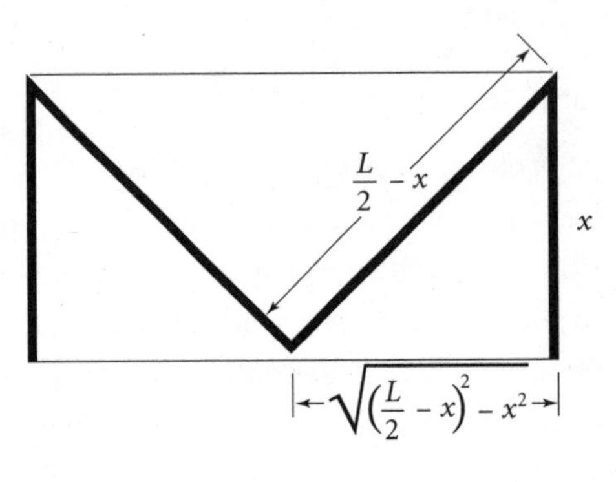

The area of the enclosing rectangle for the *M* is:

$$\text{Area} = x \cdot 2\sqrt{\left(\frac{L}{2} - x\right)^2 - x^2} = 2x\sqrt{\left(\frac{L}{x} - x\right)^2 - x^2}.$$

The form of the equation above shows the role played by *L*.

Given the same length of seed strip, students will find that the same value of x produces the maximum area for both the *M* and *S*. However, the maximum area for the *M* is larger than that for the *S*. (Using calculus, it can be shown that the maximum areas are $\frac{L^2}{6(\sqrt{3})}$ and $\frac{L^2}{12}$ respectively.)

When the "best *M*" is measured, the angles at the top corners will be 60°. Challenge students to think of a reason why a 60° angle is there. (If they cut up the flower bed and re-form it, the area is twice the area of the triangle formed by the pieces of the *M*. What triangle of fixed perimeter has the maximum area?)

Experiment 15

School Flower Beds

Name ____

Partner(s) ____

Key Concepts

Area of a rectangle = base × height

Pythagorean Theorem: $a^2 + b^2 = c^2$

a b c

Collect the Data, Initial S

Describe the experiment.

____

____

____

____

____

Data Collection

x Base	Height	Area

Points To Be Plotted

x	y Area

Enter the points as data points in your calculator, then plot them. Copy the points from the calculator display to the screen diagram below. Record the screen ranges, and label the axes.

Data Graph

Xmin = __________

Xmax = __________

Ymin = __________

Ymax = __________

Experiment 15

School Flower Beds

Mathematical Analysis, Initial S

Name ______________________________

Partner(s) ______________________________

x

Assume the strip of seeds has length L. Let x represent the base of the rectangle.

How much of the seed strip is used for the horizontal pieces of the S?

____

What is left for the vertical pieces? ______________________________

Write an equation for the area of the flower bed.

Area = ______________________________
(in terms of L and x)

For your experiment, L = ______________________________.

For your setup, what is the area function (substitute your value of L)?

Area = y = ______________________________
(in terms of x)

Graph your function (if you have a TI-82, use [Y=] and [GRAPH]; if you have a TI-81, use DrawF). Do your data points lie *close* to the graph? If not, go back and determine the source of your error. Add the graph of the function to the Data Graph on the Collect the Data sheet.

Experiment 15

School Flower Beds

Interpret Your Findings, Initial S

Name ____________________

Partner(s) ____________________

Answer the following questions. Show your work. If you have not entered the function as **Y1**, do so now and graph it.

1. Use (TRACE) to find the value of x for which the area is the largest.

   The maximum area is __________ . It occurs when $x =$ __________ .

2. What appears to be the relationship between the value of x in question 1, and L?

   ____

   ____

3. Make the S that you found in question 1 and question 2. Draw a scale model.

4. If your strip had been twice as long, what would the maximum area have been? ________

5. Which school letters—T, E, F, N, X, or U—would give a larger maximum area than the S? ________

   Which would give a smaller maximum area? __________

   Explain your reasoning.

   ____

   ____

   ____

   ____

Experiment 15

School Flower Beds

Collect the Data, Initial M

Name ______________________________

Partner(s) ______________________________

Key Concepts

Area of a rectangle = base × height

Pythagorean Theorem: $a^2 + b^2 = c^2$

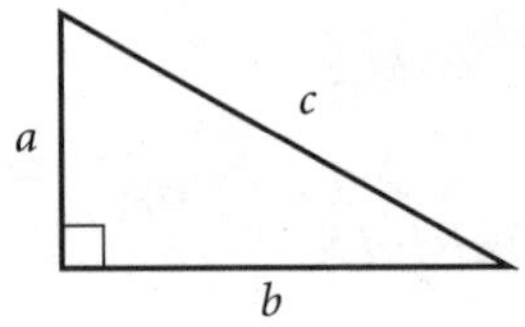

Describe the experiment.

____

____

____

____

____

____

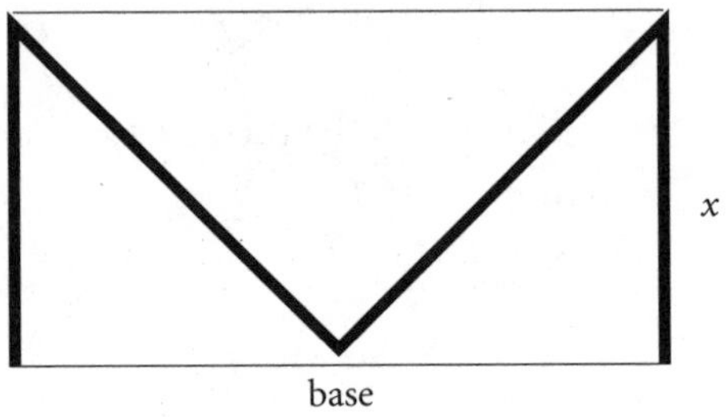

Data Collection

x Height	Base	Area

Points To Be Plotted

x	y Area

Enter the points as data points in your calculator, then plot them. Copy the points from the calculator display to the screen diagram below. Record the screen ranges, and label the axes.

Data Graph

Xmin = __________

Xmax = __________

Ymin = __________

Ymax = __________

Experiment 15

School Flower Beds

Mathematical Analysis, Initial M

Name ______________________________

Partner(s) ______________________________

x

Assume the strip has length L. Let x represent the height of the rectangle.

How much of the seed strip is used for half of the M?

(For the ⩘.) ______________________________

Therefore, how much of the seed strip is used for the hypotenuse? ________

Use the Pythagorean Theorem to solve for the other leg of the triangle formed by half the M______________________________

The entire base of the M would be ______________________________

Write an equation for the area.

Area = ______________________________

(in terms of L and x)

For your experiment, L = ______________________________

For your setup, what is the function (substitute your value of L)?

Area = y = ______________________________

(in terms of x)

Graph your function (if you have a TI-82, use (Y=) and (GRAPH); if you have a TI-81, use DrawF). Do your data points lie *close* to the graph? If not, go back and determine the source of your error. Add the graph of the function to the Data Graph on the Collect the Data sheet.

Experiment 15

School Flower Beds

Interpret Your Findings, Initial M

Name ______________________________

Partner(s) ______________________________

Answer the following questions. Show your work. If you have not entered the function as **Y1**, do so now and graph it.

1. Use (TRACE) to find the value of x for which the area is the largest.

   The maximum area is __________ . It occurs when $x =$ __________ .

2. What appears to be the relationship between the value of x in question 1, and L?

   ____

   ____

3. Make the M that you found in question 1 and question 2. Draw a scale model and measure all the angles.

4. If your strip had been twice as long, what would the maximum area have been? __________

5. Which school letters—T, E, F, N, X, or U—would give a larger maximum area than the M ________

   Which would give a smaller maximum area? __________

   Explain your reasoning.

   ____

   ____

   ____

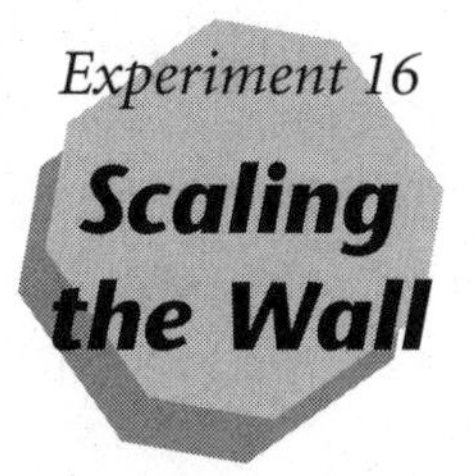

Teaching Notes

In this experiment, students model a building that is surrounded by a barrier. Students lean a ramp against the barrier so that it just reaches the building, then measure the length of the ramp and the height the ramp reaches on the building. They repeat the procedure with ramps of different lengths. The length of the ramp is the *independent variable,* and the height at which a ramp of length x touches the building is the *dependent variable.*

Key Concepts

Pythagorean Theorem: $a^2 + b^2 = c^2$

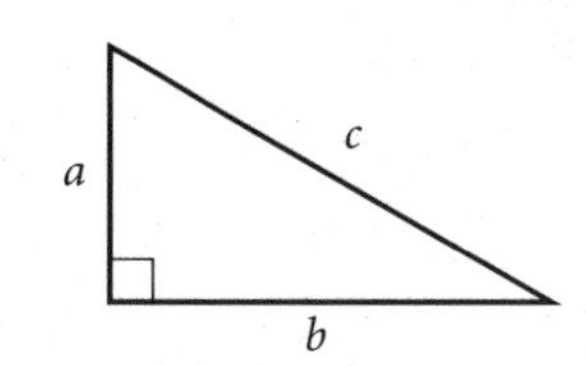

Parametric equations

Similar triangles: Corresponding sides are proportional

Equipment

several boards of various lengths, 15"–36"

A collection of boards can serve several groups of students.

piles of blocks or books for the barrier, 1 per group

yardsticks or meter sticks, 1 per group

Procedure

Note: To graph the function found on the Mathematical Analysis sheet, students must be able to graph equations of the form $x = g(y)$. This is done in the parametric mode by graphing the equations:

$$X_{1T} = \text{expression for } g(Y_{1T}); Y_{1T} = T$$

Tmin, Tmax are the same as Ymin, Ymax. Start with Tstep = 0.1, and adjust if necessary.

Each group sets up its fixed barrier, noting height and position: The distance d of the barrier from the wall and its height h are constants that will affect the value of y. Students lean an assortment of ramps against the wall. The ramps must touch the fixed barrier.

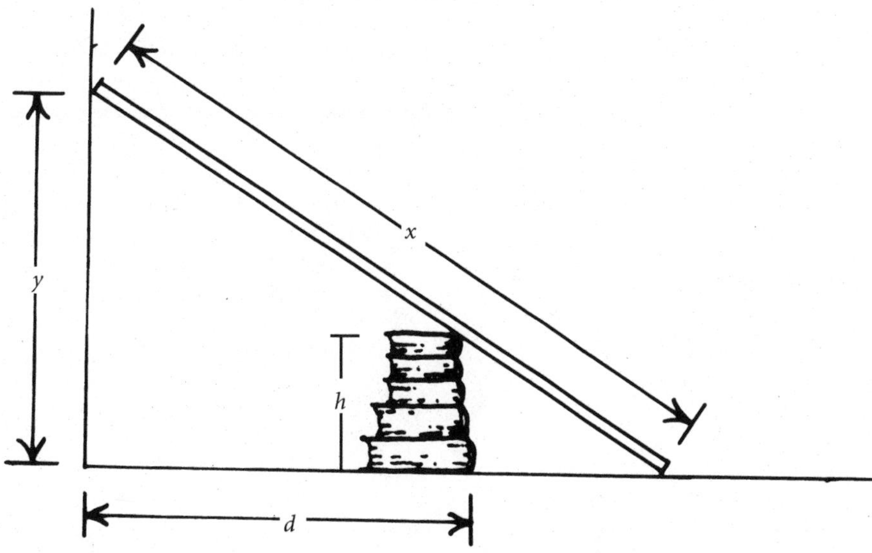

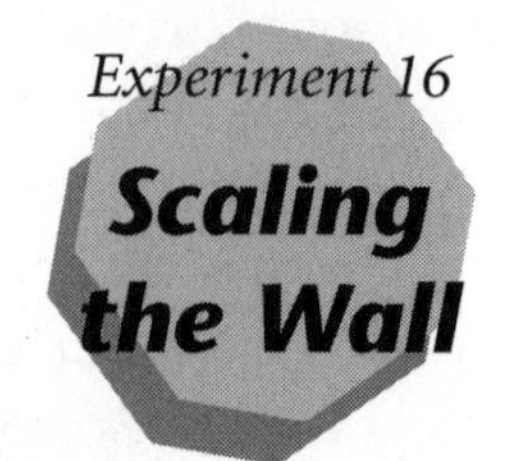

Teaching Notes, page 2

This problem has several complications. Students may set up the barrier in such a way that some ramps will not reach across it to the wall; there will be no data for these shorter ramps. For the ramps that do reach the wall, students will record two values of y: a "high y" value at the steepest incline, and a "low y" value at the flattest incline. Thus, each value of x will generate two data points.

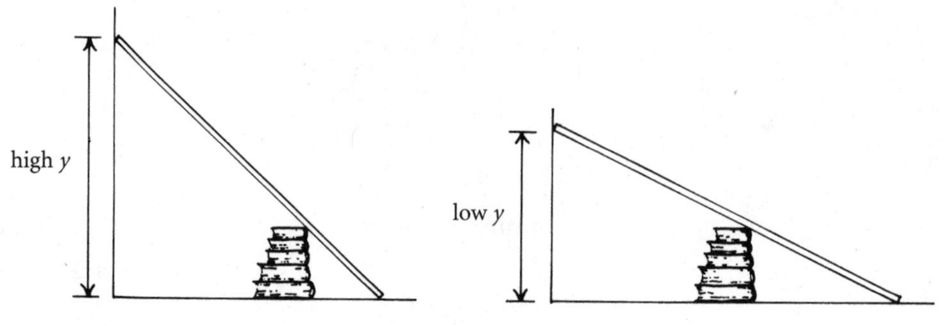

Consider the following sample data: $d = 9$, $h = 2.5$

x	**high y**	**low y**
20	16.75	4.75
24	21.75	4.25
16.875	12.5	6
15.25	8.5	8
18	14.75	5.5
36	34.5	3.5
15	did not reach	

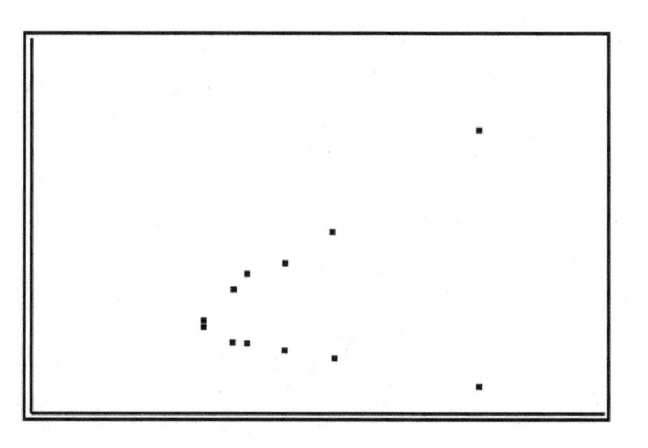

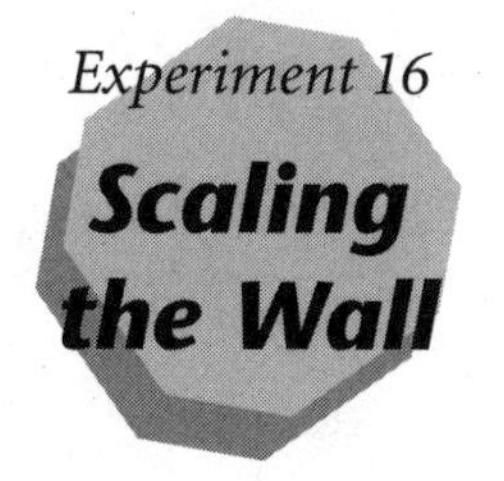

Teaching Notes, page 3

Using the following diagrams, the Pythagorean Theorem, and similar triangles, we obtain the following:

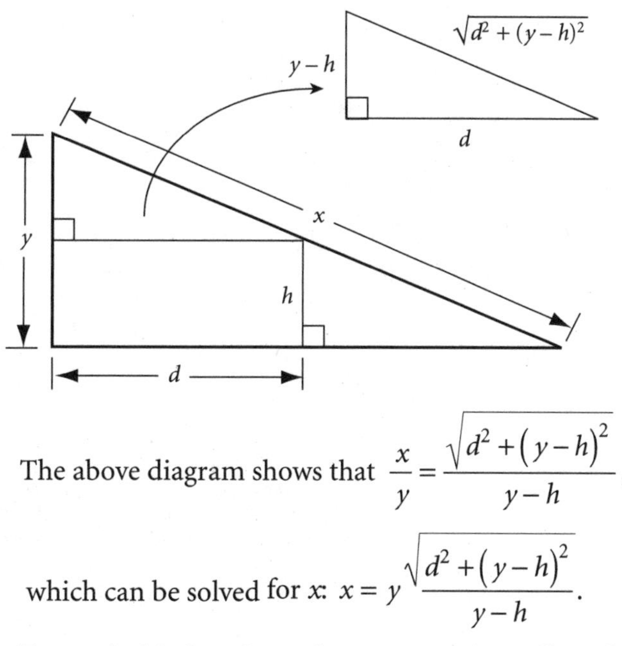

The above diagram shows that $\dfrac{x}{y} = \dfrac{\sqrt{d^2 + (y-h)^2}}{y-h}$,

which can be solved for x: $x = y\dfrac{\sqrt{d^2 + (y-h)^2}}{y-h}$.

To graph this function, select parametric mode and enter the following functions:

$$X_{1T} = Y_{1T}\frac{\sqrt{d^2 + (Y_{1T}-h)^2}}{Y_{1T}-h};\ Y_{1T} = T$$

For the sample data above, screen ranges of Xmin = 0, Xmax = 50, Ymin = 0, and Ymax = 50 will show the data. Since T corresponds to $y = Y_{1T}$, set Tmin = 0, Tmax = 50. A Tstep of 0.2 is sufficiently small. To see the function, redraw the scatter plot. Since the data points are plotted after the curve is drawn, they will be invisible if they lie on top of the graph.

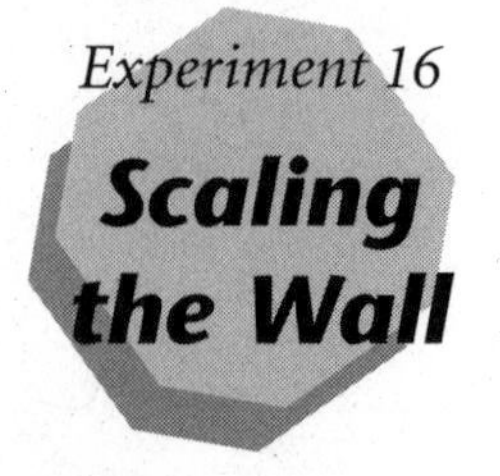

Collect the Data

Name ______________________________

Partner(s) ______________________________

Key Concepts

Pythagorean Theorem: $a^2 + b^2 = c^2$

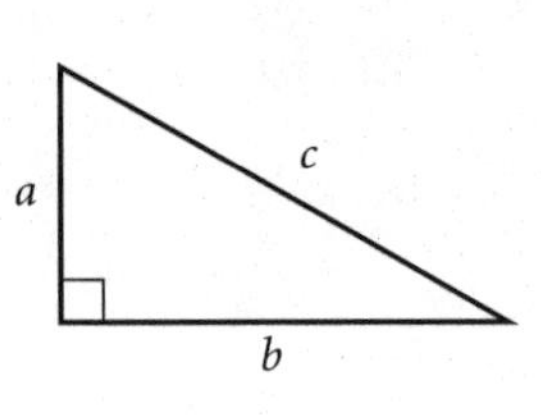

Parametric equations

Similar triangles: Corresponding sides are proportional

Describe the experiment.

____

____

____

____

____

Data Collection

x	High y	Low y

Points To Be Plotted

x	y

x	y

Enter the points as data points in your calculator, then plot them. Copy the points from the calculator display to the screen diagram below. Record the screen ranges, and label the axes.

Data Graph

Xmin = ______________

Xmax = ______________

Ymin = ______________

Ymax = ______________

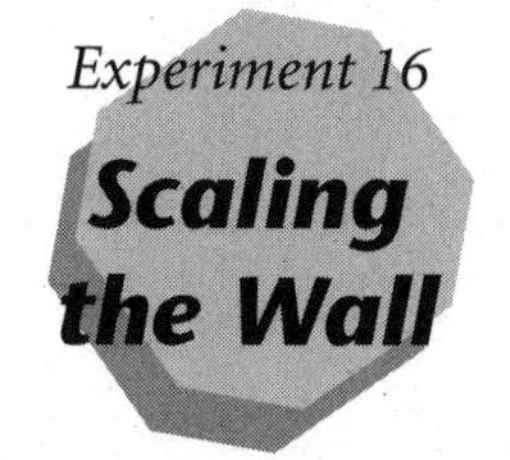

Name ____

Partner(s) ______________________________________

Mathematical Analysis

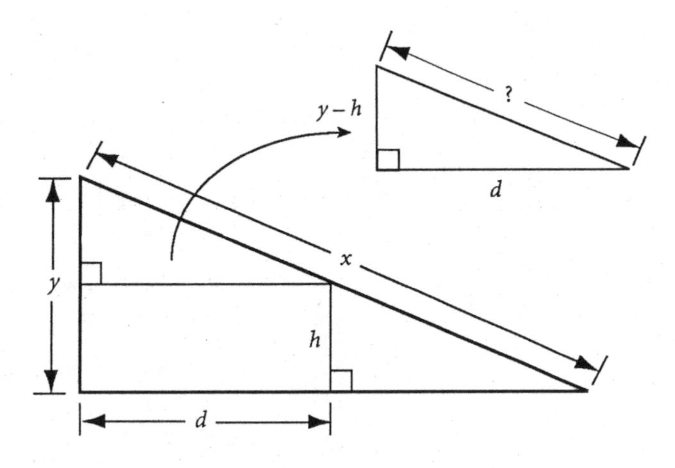

Assume the barrier has height h and is a distance d from the wall. Using the Pythagorean Theorem, find the hypotenuse (?) in terms of d, y, and h.

____

Using similar triangles, find the corresponding sides to complete the ratio:

$\frac{x}{y} =$

Solve for x: $x =$ ____________________________________
(in terms of y, h, and d)

For your experiment, $h =$ ______________ and $d =$ ______________ .

For your setup, what is the function (substitute your values of h and d)?

$x =$ ____
(in terms of y)

Using the parametric mode, graph your function. Do your data points lie *close* to the graph? If not, go back and determine the source of your error. Add the graph of the function to the Data Graph on the Collect the Data sheet.

Experiment 16

Scaling the Wall

Interpret Your Findings

Name ______________________________

Partner(s) ______________________________

Answer the following questions. Show your work.

1. Use (TRACE) or move the cursor around the screen to find the value of the smallest value of x. (If (TRACE) does not work on your calculator, use the arrow keys to move the cursor). ______________________________

   Describe the physical significance of this value.

   ____

   ____

2. For the value of x in question 1, what is the distance of the foot of the ramp from the wall? _______

   Show how you found this distance.

   ____

   ____

   ____

3. What are the lengths of the ramps that would exactly reach a height of 10?

   ____

   ____

4. What are the lengths of the ramps that would exactly reach a height of 20?

   ____

   ____

5. Had your barrier been 4 units higher, what would the length of the shortest possible ramp have been? ______

   Explain in detail how you obtained your answer.

   ____

   ____

   ____

Teaching Notes

In this experiment, students simulate the taking of a class photograph. Geometrically, it is an investigation of the distance between points on a fixed line segment and a fixed point that is not on the line segment (the camera). The *independent variable* is the distance of a point (a student) from the left endpoint of the segment, and the *dependent variable* is that student's distance from the camera.

Key Concept

Law of cosines: $c^2 = a^2 + b^2 - 2ab \cos x$

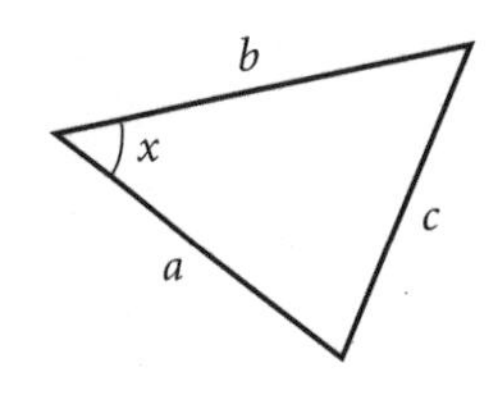

Equipment

graph paper (see page 131), 1 sheet per group

rulers, 1 per group

markers, 2 colors per group

clear protractor, 1 per group

You may want to make and use circular protractors by copying the blackline master on page 132 onto overhead-transparency sheets. Mark the center dot and the 0-degree line in color.

Procedure

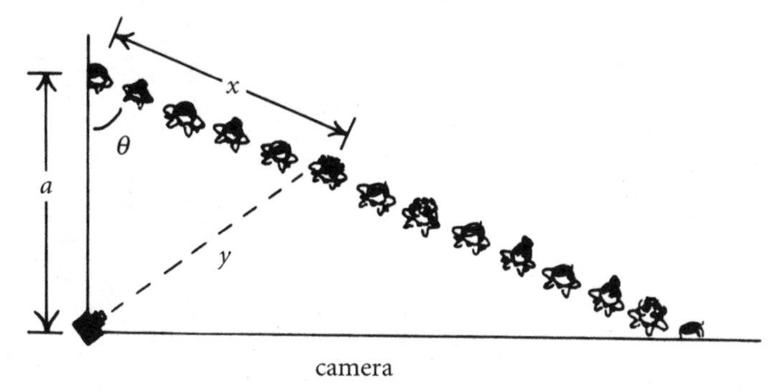

Explain that students will be modeling a situation in which a camera is placed in a corner of the room and students stand in a straight line across the room to have a class photograph taken. For focusing purposes, it is necessary to determine the distance of each student from the camera, which will be a function of the student's location in line.

Have each group draw a corner and a line on their graph paper to simulate the situation. Each group should use a different value for a and for θ, with the value of θ being at least 50°. Students measure the distance to the camera for about six different "students" in the line; that is, for six different values of x.

As students complete the Interpret Your Findings sheet, they will discover that the line from the camera to the closest student is perpendicular to the line along which the class is standing.

Teaching Notes, page 2

Mathematical Analysis

Using the law of cosines and solving for y:

$$y^2 = a^2 + x^2 - 2ax \cos \theta$$

$$y = \sqrt{a^2 + x^2 - 2ax \cos \theta}$$

The final question on the Interpret Your Findings sheet asks students to compare the distance from the camera of Henri, who stands halfway between Nigel and Frieda, to the average of Nigel's and Frieda's distances from the camera. Geometrically this can be solved by constructing midpoints and circles:

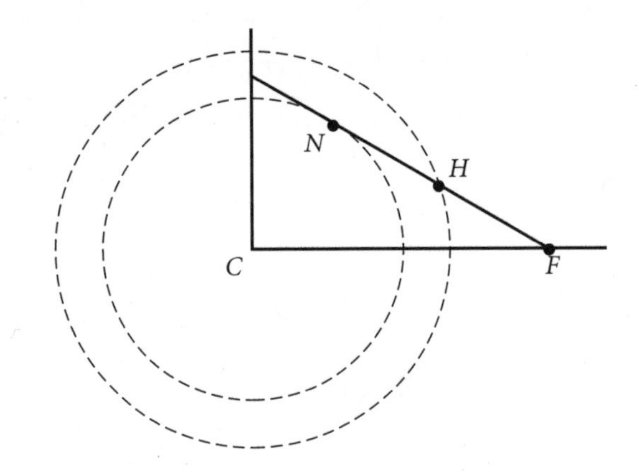

Experiment 17

Class Photo

Name ______________________________

Partner(s) ______________________________

Key Concept

Law of cosines: $c^2 = a^2 + b^2 - 2ab \cos x$

b
x
c
a

Collect the Data

Describe the experiment.

____

____

____

____

____

Data Collection

Distance from End of Line	Distance from Camera

Points To Be Plotted

x	y

Enter the points as data points in your calculator, then plot them. Copy the points from the calculator display to the screen diagram below. Record the screen ranges, and label the axes.

Data Graph

Xmin = ____________

Xmax = ____________

Ymin = ____________

Ymax = ____________

Experiment 17

Class Photo

Name ______________________________

Partner(s) ______________________________

Mathematical Analysis

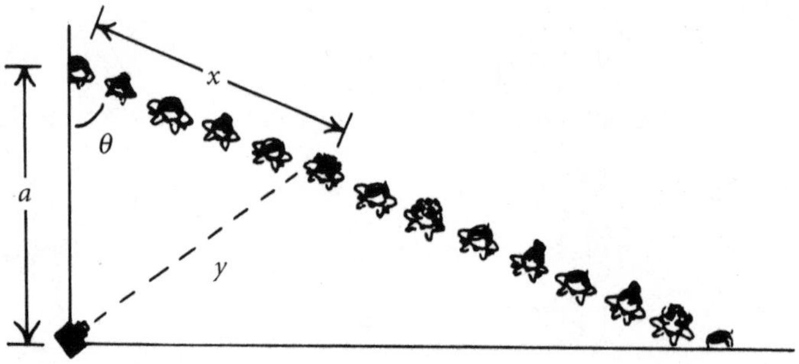

x
θ
a
y
camera

Write an expression for y^2 using the law of cosines.

$y^2 =$ ______________________________

Simplify this expression.

$y =$ ______________________________

In your experiment, what was the value of a?

$a =$ ______________________________

$\theta =$ ______________ and $\cos \theta =$ ______________

Substitute these values into your equation.

$y =$ ______________________________

Graph your function (if you have a TI-82, use (Y=) and (GRAPH); if you have a TI-81, use DrawF). Do your data points lie *close* to the graph? If not, go back and determine the source of your error. Add the graph of the function to the Data Graph on the Collect the Data sheet.

Name ______________________________

Partner(s) ______________________________

Answer the following questions. Show your work. If you have not entered the function as **Y1**, do so now and graph it.

Interpret Your Findings

1. Use (TRACE) and (ZOOM) to find where the student who wishes to be as close as possible to the camera would stand.

   The closest person would be ______ units from the left end of the line.

   The minimum distance from the camera is ___________ .

2. Use (TRACE) and (ZOOM) to find where the student who wishes to be as far as possible from the camera would stand.

   The farthest person would be _____ units from the left end of the line.

   The maximum distance from the camera is ____________________ .

3. Make a scale drawing of your triangle, labelling the points as in the example below. Mark the locations of the nearest student, Nigel, and of the farthest student, Frieda. Draw a line from the camera to Frieda. Measure the angles determined by the camera (point *C*), Nigel, and Frieda.

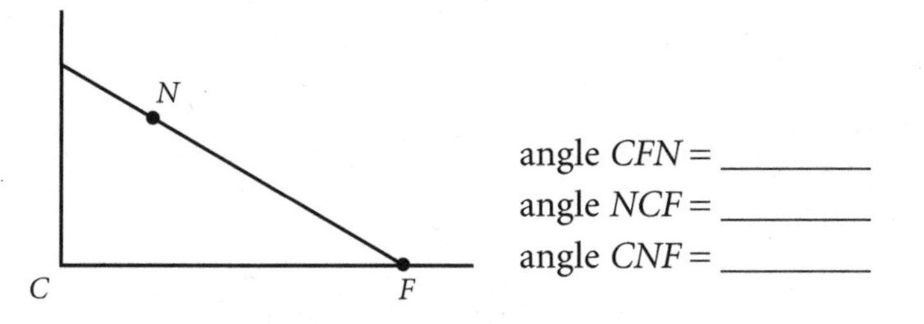

   angle CFN = ________
   angle NCF = ________
   angle CNF = ________

   Can you explain your results for angle CNF?

   ____

4. Find the average of the camera's distances to Nigel and Frieda. ______

5. Suppose the camera is focused at that distance. Let Henri be on the line, halfway between Nigel and Frieda. How far is Henri from the camera?

   ____

   Compare Henri's distance from the camera to the average of the camera's distances to Nigel and Frieda.

   ____

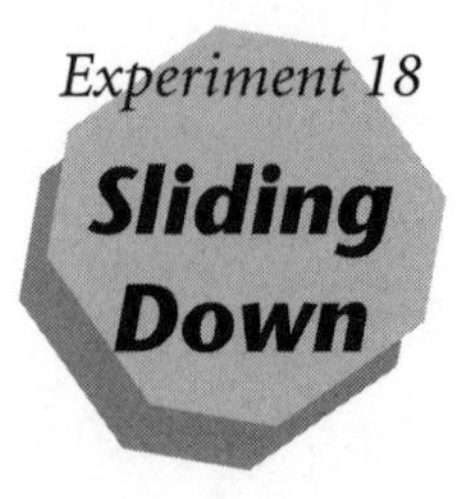

This experiment is an investigation of what happens to a person standing on a ramp as the ramp slides down the wall. The distance from the base of the ramp to the base of the wall is the *independent variable,* and the distance from the person on the ramp to the corner of the base and the wall is the *dependent variable.*

Key Concepts

Distance formula: distance = $\sqrt{(x_2 - x_1)^2 + (y_2 - y_1)^2}$

Pythagorean Theorem: $a^2 + b^2 = c^2$

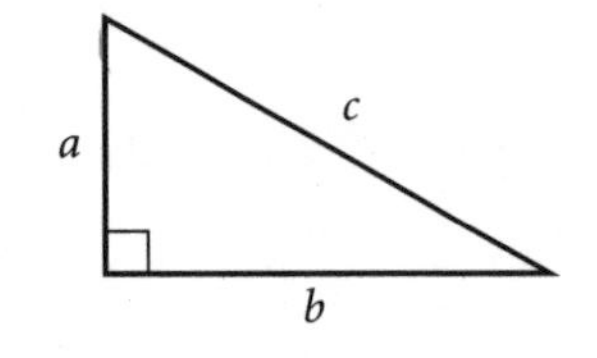

Equipment

boards of varying lengths, 12"–24", 1 per group

pushpins, 2 per group

yardsticks or meter sticks, 1 per group

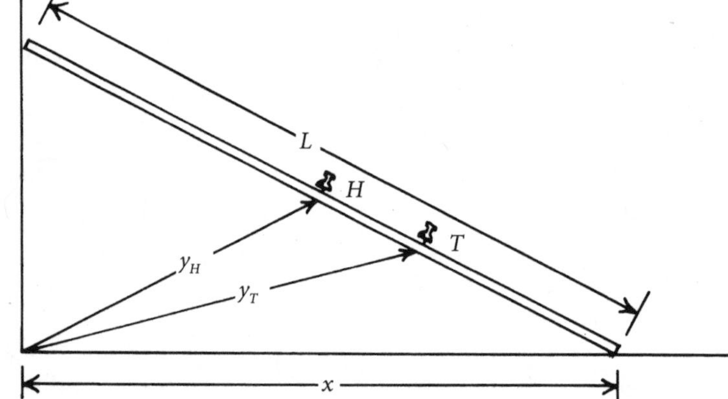

Procedure

Have students imagine two people standing on the ramp as it slides down the wall. "Helene" is halfway up the ramp. "Thelma" is one-third of the way up the ramp. Students use pushpins to mark these positions on the ramp.

Students lean the ramp against the wall. They measure the distance from the base of the ramp to the base of the wall and record the value as the independent variable x. The dependent variable is the distance from the pushpin to the corner of the floor and the wall. Varying x, students measure the distances of Helene's pushpin and Thelma's pushpin from the corner.

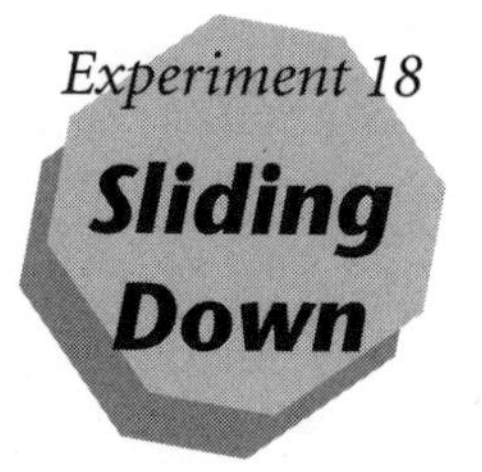

Teaching Notes, page 2

Mathematical Analysis

Students should find that Helene is always a fixed distance from the corner. The mathematical analysis is done using coordinate geometry and the distance formula. You might challenge students to find a geometric proof for this fact.

Helene's coordinates, (Hx, Hy), are found by the following process:

The vertical leg of the right triangle is $\sqrt{L^2 - x^2}$.

Therefore, the coordinates of H are $\left(\frac{x}{2}, \sqrt{\frac{L^2 - x^2}{2}}\right)$, which can also be written as $\left(\frac{x}{2}, \sqrt{\left(\frac{L}{2}\right)^2 - \left(\frac{x}{2}\right)^2}\right)$.

Using the distance formula, we arrive at:

$$y_H = \sqrt{\left(\frac{x}{2}\right)^2 + \left(\frac{L}{2}\right)^2 - \left(\frac{x}{2}\right)^2}.$$

$$y_H = \sqrt{\left(\frac{L}{2}\right)^2}$$

$$y_H = \frac{L}{2}$$

This coincides with the geometry theorem that states that the midpoint of the hypotenuse is equidistant from all three vertices.

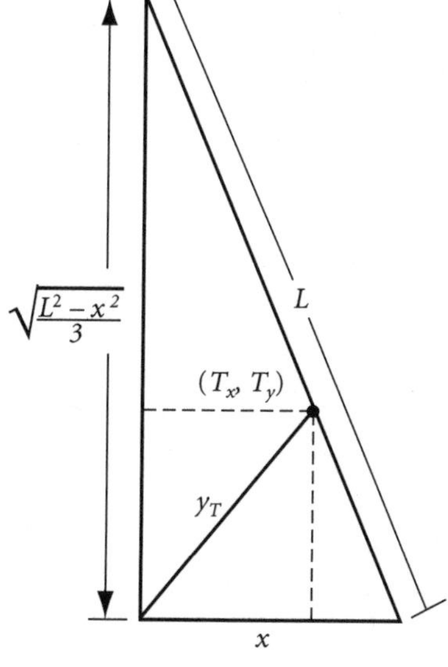

Thelma's coordinates, (Tx, Ty), are found by the following process:

The vertical leg of the right triangle is $\sqrt{L^2 - x^2}$. Therefore, the coordinates of T are $\left(\frac{2}{3}x, \frac{\sqrt{L^2 - x^2}}{3}\right)$.

Using the distance formula, we arrive at:

$$y_T = \sqrt{\left(\frac{2}{3}x\right)^2 + \left(\frac{\sqrt{L^2 - x^2}}{3}\right)^2}$$

$$y_T = \sqrt{\frac{4x^2}{9} + \frac{L^2}{9} - \frac{x^2}{9}}$$

$$y_T = \sqrt{\frac{3x^2}{9} + \frac{L^2}{9}}$$

$$y_T = \frac{\sqrt{3x^2 + L^2}}{3}$$

Experiment 18

Sliding Down

Name ______________________________

Partner(s) ______________________________

Key Concepts

Distance formula: distance $= \sqrt{(x_2 - x_1)^2 + (y_2 - y_1)^2}$

Pythagorean Theorem: $a^2 + b^2 = c^2$

Collect the Data

Describe the experiment.

____

____

____

____

____

Data Collection

x	Helene's Distance	Thelma's Distance

Points To Be Plotted

x	y_H	y_T

For Helene's position: What would the graph of the points (x, Helene's distance) look like?

____

For Thelma's position: Enter the points as data points in your calculator, then plot them. Copy the points from the calculator display to the screen diagram below. Record the screen ranges, and label the axes.

Data Graph

Xmin = ________

Xmax = ________

Ymin = ________

Ymax = ________

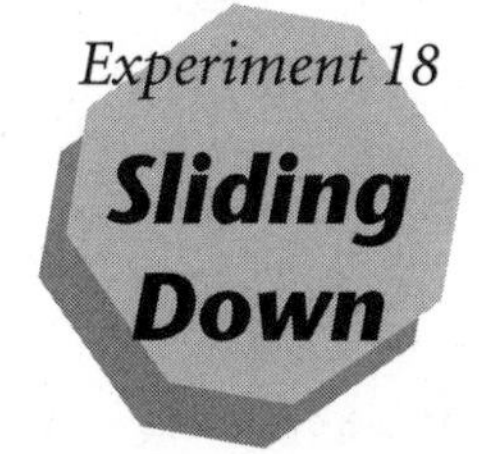

Name ______________________________

Partner(s) ______________________________

Mathematical Analysis, Helene

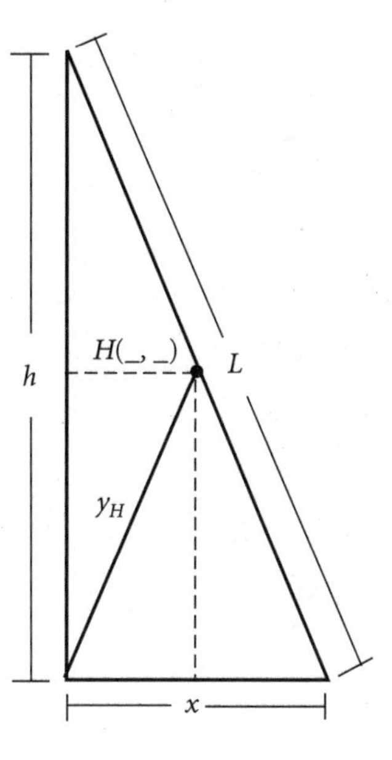

Assume the ramp has length L. What are the coordinates of Helene's position? (Helene stands halfway up the ramp.)

$H =$ (_______ , _______)
(in terms of x and h)

Use the Pythagorean Theorem to write an equation for h.

$h =$ ______________________________
(in terms of x and L)

Rewrite the coordinates.

$H =$ (_______ , _______________)
(in terms of x and L)

Using the distance formula, find an equation for Helene's distance from the corner.

distance = ______________________________

Simplify it.

distance = ______________________________

What is your value of L? ________ Using your value of L, rewrite y as a function of x.

distance = y_H = ______________________________

Graph your function (if you have a TI-82, use (Y=) and (GRAPH); if you have a TI-81, use DrawF). Do your data points lie *close* to the graph? If not, go back and determine the source of your error. Add the graph of the function to the Data Graph on the Collect the Data sheet.

Experiment 18

Sliding Down

Interpret Your Findings, Helene

Name ______________________________

Partner(s) ______________________________

Answer the following questions. Show your work. If you have not entered the function as **Y1**, do so now and graph it.

1. Do your results agree with what you know about the relationship of the midpoint of the hypotenuse to the vertices? Explain.

____

____

2. Look at the path Helene traveled as the ramp slid down. What is the description of the locus of these points? ______________
How does your data predict this locus?

____

____

____

3. Identify the graphs of the distance from the corner as a function of x, for Bill (at the bottom), Tip (at the top), and Helene (at the midpoint), as the 10-foot ramp they are standing on slides down.

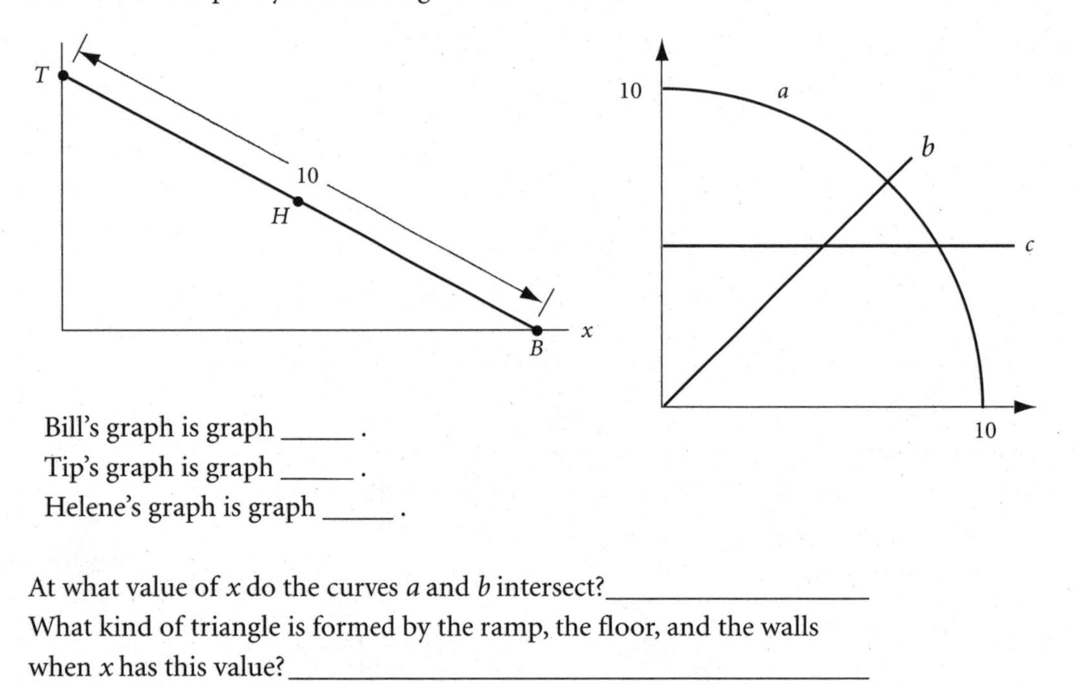

Bill's graph is graph _____ .
Tip's graph is graph _____ .
Helene's graph is graph _____ .

At what value of x do the curves a and b intersect?______________
What kind of triangle is formed by the ramp, the floor, and the walls when x has this value?______________

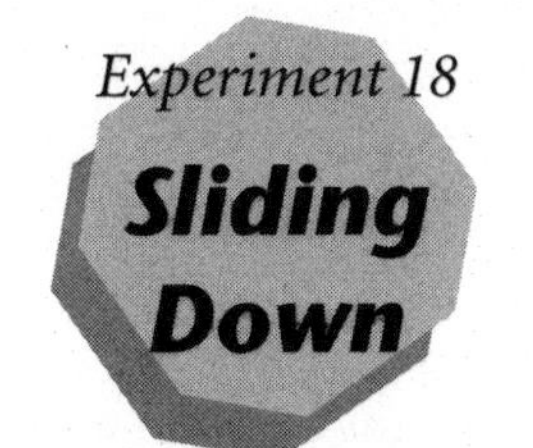

Name ____________________

Partner(s) ____________________

Mathematical Analysis, Thelma

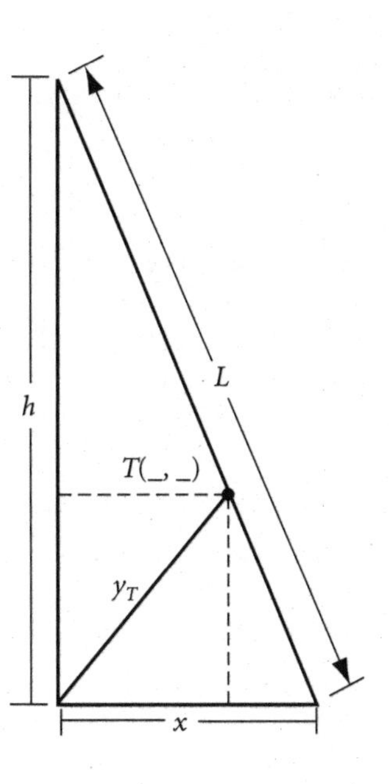

Assume the ramp has length L. What are the coordinates of Thelma's position? (Thelma stands one-third of the way up the ramp.)

$T =$ (________ , ________)
(in terms of x and h)

Use the Pythagorean Theorem to write an equation for h.

$h =$ ____________________
(in terms of x and L)

Rewrite the coordinates.
$T =$ (________ , ________)
(in terms of x and L)

Using the distance formula, find an equation for Thelma's distance from the corner.
distance = ____________________

Simplify it.

distance = ____________________

What is your value of L? ________ Using your value of L, rewrite y as a function of x.

distance = y_T = ____________________

Graph your function (if you have a TI-82, use (Y=) and (GRAPH); if you have a TI-81, use DrawF). Do your data points lie *close* to the graph? If not, go back and determine the source of your error. Add the graph of the function to the Data Graph on the Collect the Data sheet.

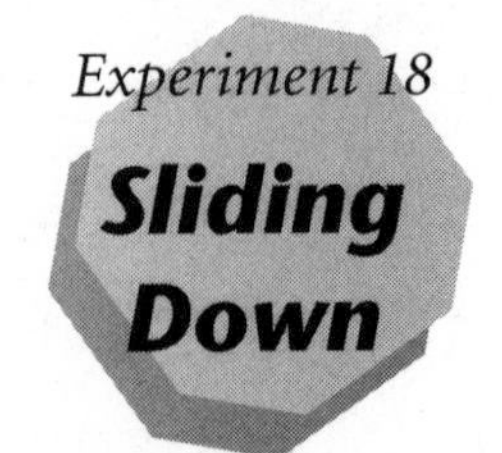

Interpret Your Findings, Thelma

Name ______________________________

Partner(s) ______________________________

Answer the following questions. Show your work. If you have not entered the function as **Y1**, do so now and graph it.

1. Use (TRACE) and (ZOOM) to find the value of the y-intercept. ________

   What actual physical situation does that point represent?

   ____

   ____

   ____

2. Use (TRACE) and (ZOOM) to find the maximum distance Thelma can be from the origin. ____________

3. Investigate the shape of Thelma's path as the ramp slides down.

   When $x = 0$, Thelma's position is (____ , ____).

   When $x = L$, Thelma's position is (____ , ____).

   Plot these points. Determine Thelma's path between the points, and sketch it.

   Explain how you found the path.

   ____

   ____

   ____

Teaching Notes

This is the first of two experiments involving parallelograms made from cut-up boxes. Both experiments involve trigonometry. In this first experiment, the *independent variable* is the measure of one of the base angles, and the *dependent variable* is the area of the parallelogram.

Key Concepts

Area of a parallelogram = base × height

$\sin\ \theta = \dfrac{y}{r}$

Equipment

cereal boxes or other boxes, 1 per group

Make parallelogram frames by removing the tops and bottoms of the boxes and cutting the sides down to 1″–2″.

graph paper (see page 131) with cardboard backing, 1 sheet per group

pushpins (to anchor the base), 2 per group

clear circular protractors, 1 per group

You may want to make the circular protractors by copying the blackline master on page 132 onto overhead-transparency sheets. Mark the center dot and the 0-degree line in color.

transparency of the graph paper

overhead projector

Procedure

Demonstrate the experiment at the overhead projector. Draw a horizontal base line in the middle of the graph paper. Have a student hold the base of the box in place while you squash the box.

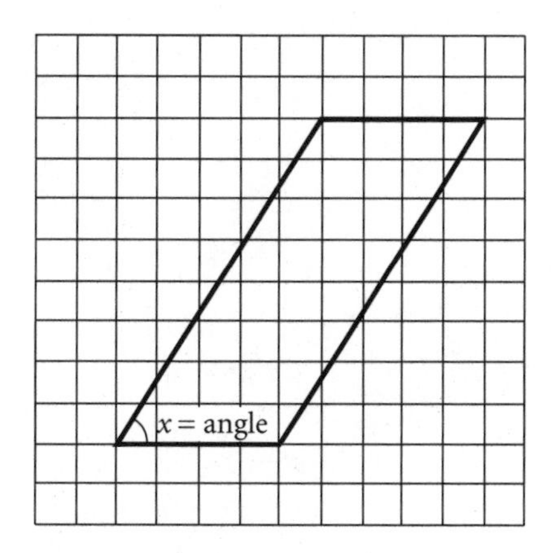

Teaching Notes, page 2

Working in groups, students measure the height of the parallelogram and then compute the area for different values of the angle x. Students will often stop measuring x at 90°. As you walk around, ask whether $x > 90$ makes sense as a value. Encourage students to use at least two values of $x > 90$ so that when the points are plotted, they can see the area grow, then diminish.

Mathematical Analysis

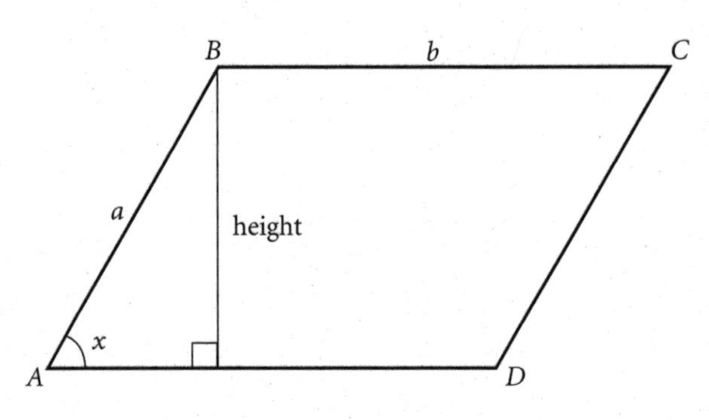

Given that x is the measure of angle A, a is the length of sides AB and CD, and b is the length of the other two sides, area = $b \cdot a \cdot \sin x$.

Experiment 19

Squashed Boxes I

Name ______________________________

Partner(s) ______________________________

Key Concepts

Area of a parallelogram = base × height

$\sin\,\theta = \dfrac{y}{r}$

r
y
θ
x

Collect the Data

Describe the experiment.

____

____

____

____

____

Data Collection

x	Height	Area

Points To Be Plotted

x	y

Enter the points as data points in your calculator, then plot them. Copy the points from the calculator display to the screen diagram below. Record the screen ranges, and label the axes.

Data Graph

Xmin = __________

Xmax = __________

Ymin = __________

Ymax = __________

Experiment 19

Squashed Boxes I

Name ______________________________

Partner(s) ______________________________

Mathematical Analysis

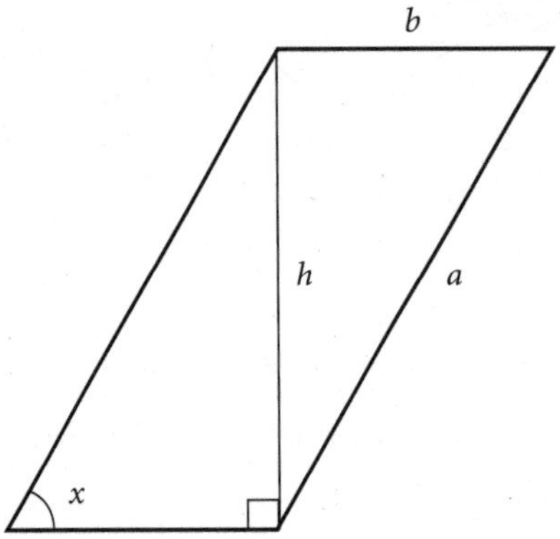

Assume the box has base b and side a. Find an expression for h.

$h =$ ______________________________

(in terms of a and x)

Write an equation for the area of the parallelogram using a, b, and x.

$y =$ area = ______________________________

(in terms of a, b, and x)

For your experiment, $b =$ ____________ and $a =$ ____________ .

For your setup, what is the function (substitute your values of b and a)?

$y =$ area = ______________________________

(function of x)

Graph your function (if you have a TI-82, use (Y=) and (GRAPH); if you have a TI-81, use DrawF). Do your data points lie *close* to the graph? If not, go back and determine the source of your error. Add the graph of the function to the Data Graph on the Collect the Data sheet.

Name ______________________________

Partner(s) ______________________________

Answer the following questions. Show your work. If you have not entered the function as **Y1**, do so now and graph it.

Interpret Your Findings

1. Use (TRACE) to find the value of x for which the area is as large as possible.

   The maximum area = __________ . It occurs for x = __________ .

2. On the back of your graph paper, make a scale drawing for the value of x you found in question 1. Draw in the diagonals, and measure the angles where they meet.

3. Use (TRACE) to find the value(s) of x for which the area is exactly half the maximum area. What is the height at this value?

4. If you had done the experiment with the box frame turned the other way (with base a and side b), what would have been the equation for the area?

   ____

5. What is the perimeter of your parallelograms? __________
   Suppose another team's parallelograms had the same perimeter as yours, but b and a were different. Would their function have been the same? ________ Would the maximum area have been the same? ________
   Explain your reasoning.

   ____

   ____

6. Suppose the perimeter is fixed, but a and b vary. If the total perimeter is 20 units, find a as a function of b. Then find an expression for the area.

   a = ____________________ (in terms of b)   Area = ____________________ (in terms of b and x)

   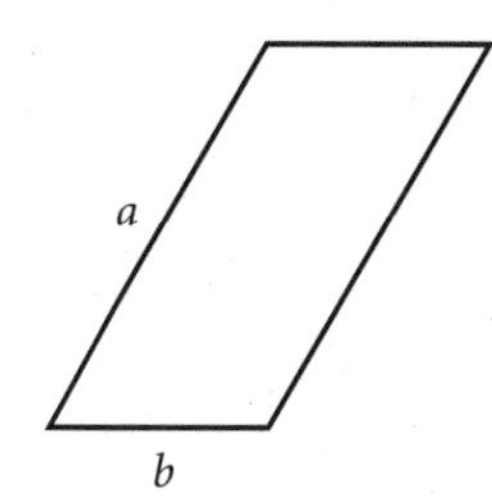

   Maximum area = ____________________ (in terms of b)

   Investigate: What value of b gives the largest area when the perimeter is 20 units?

Teaching Notes

This is the second of two experiments involving parallelograms made from cut-up boxes. The *independent variable* is the measure of one of the base angles of the parallelogram, and the *dependent variable* is the length of the opposite diagonal.

Key Concept

Law of cosines: $c^2 = a^2 + b^2 - 2ab \cos x$

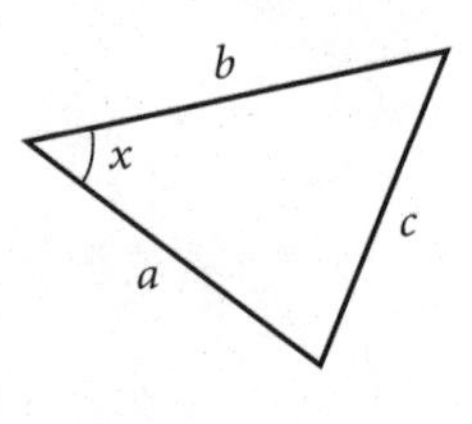

Equipment

cardboard parallelogram frames from Experiment 19, 1 per group

lengths of elastic thread, 1 per group

rulers, 1 per group

transparent ruler

overhead projector

clear protractor, 1 per group

You may want to make and use the circular protractors by copying the blackline master on page 132 onto overhead-transparency sheets. Mark the center dot and the 0-degree line in color.

scissors or craft knives

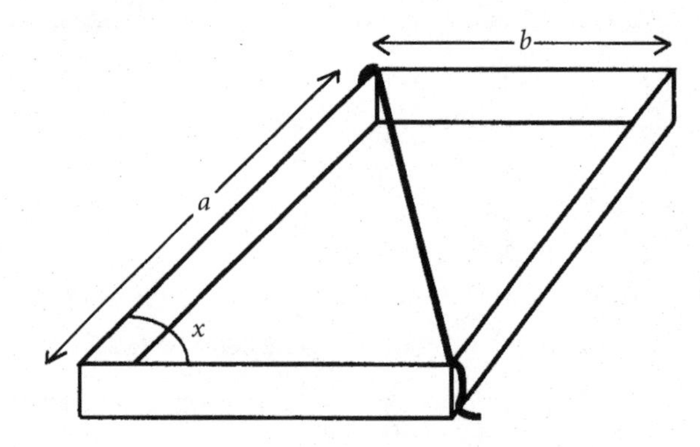

Procedure

Direct students to make a diagonal on their frame using elastic thread. They cut a one-quarter-inch slit in two opposite corners of their frame, then stretch a length of thread, longer than half the perimeter of the frame, across the diagonal. They adjust the tension so the elastic is taut but has some give.

Model the experiment at the overhead projector using a frame, protractor, and transparent ruler. Place the frame on the projector, and position the protractor under an elastic-free corner. Squash the box, then measure the angle, x, and the length of the diagonal, y. Repeat the procedure for another value of x. Point out that it will be necessary to adjust the tension of the elastic for different values of x.

Teaching Notes, page 2

Have students include both acute angles and obtuse angles as they collect their data.

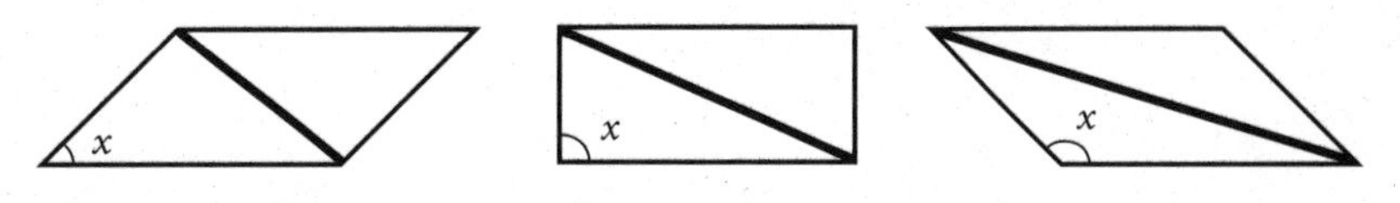

Mathematical Analysis

By the law of cosines:

Law of cosines: $y^2 = a^2 + b^2 - 2ab \cos x$

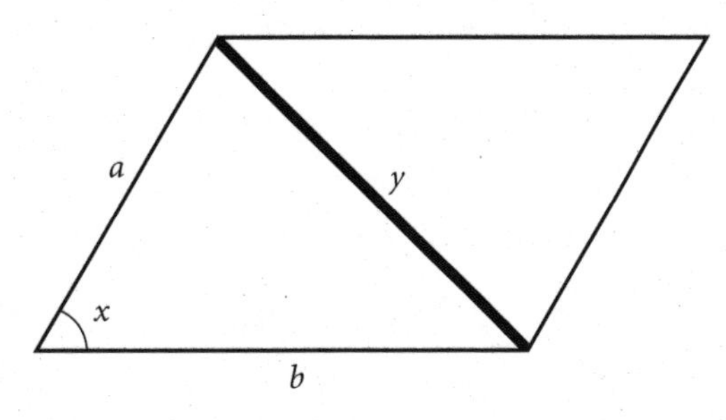

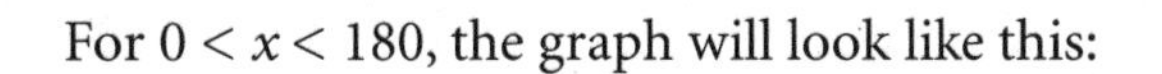

For $0 < x < 180$, the graph will look like this:

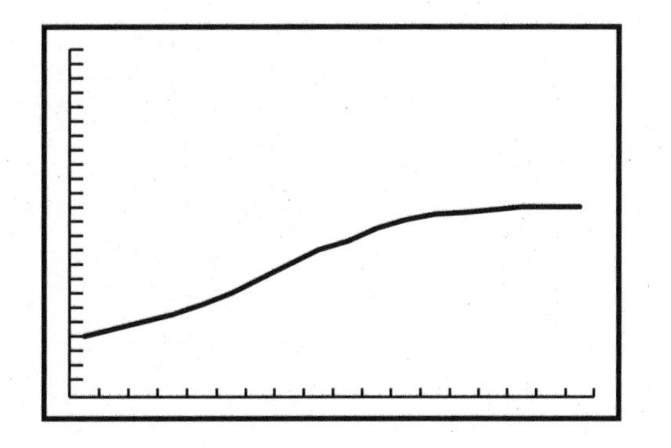

Law of cosines: $c^2 = a^2 + b^2 - 2ab \cos x$

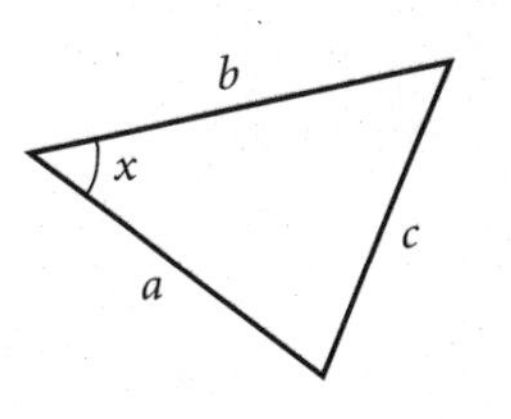

Experiment 20

Squashed Boxes II

Name ______________________________

Partner(s) ______________________________

Key Concept

Law of cosines: $c^2 = a^2 + b^2 - 2ab \cos x$

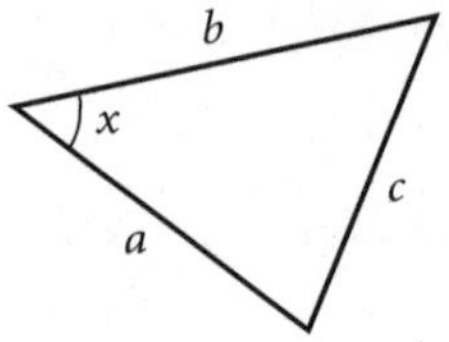

Collect the Data

Describe the experiment.

____

____

____

____

____

Data Collection

Angle	Diagonal

Points to Be Plotted

x Angle	y Diagonal

Enter the points as data points in your calculator, then plot them. Copy the points from the calculator display to the screen diagram below. Record the screen ranges, and label the axes.

Data Graph

Xmin = ____________

Xmax = ____________

Ymin = ____________

Ymax = ____________

Experiment 20

Squashed Boxes II

Name ______________________________

Partner(s) ______________________________

Mathematical Analysis

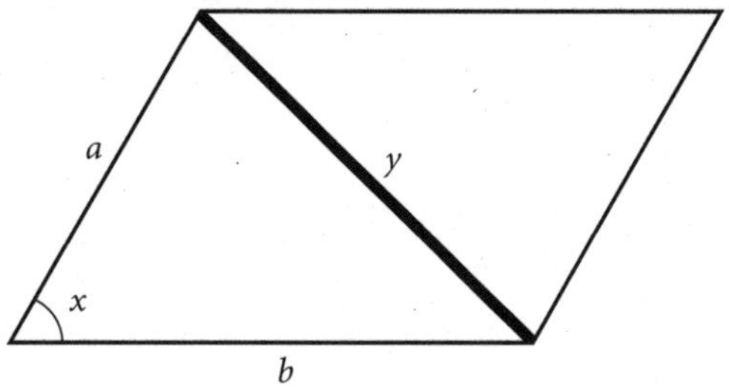

Assume the box has base b and side a. Find an equation for y^2

$y^2 =$ ______________________________
(in terms of a, b, and x)

For your experiment, $b =$ ____________ and $a =$ ____________ .

For your setup, what is the function (substitute your values of b and a)?

$y =$ ______________________________
(in terms of x)

Graph your function (if you have a TI-82, use (Y=) and (GRAPH); if you have a TI-81, use DrawF). Do your data points lie *close* to the graph? If not, go back and determine the source of your error. Add the graph of the function to the Data Graph on the Collect the Data sheet.

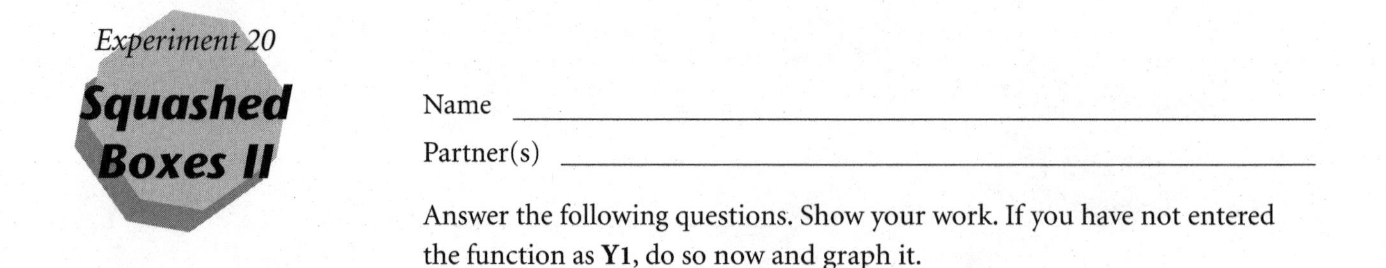

Name ______________________________

Partner(s) ______________________________

Answer the following questions. Show your work. If you have not entered the function as **Y1**, do so now and graph it.

Interpret Your Findings

1. Suppose for your parallelogram b was the same, but $a = 2b$ (twice as long). What effect will this have on your graph?

____

____

2. Suppose for your parallelogram b was the same, but $a = \frac{b}{2}$. Explain the effect on your curve.

____

____

3. What values of a and b, if any, would make this curve go through (0, 0)?

____

____

4. Using your original values for a and b, write the equation for the other diagonal.

____

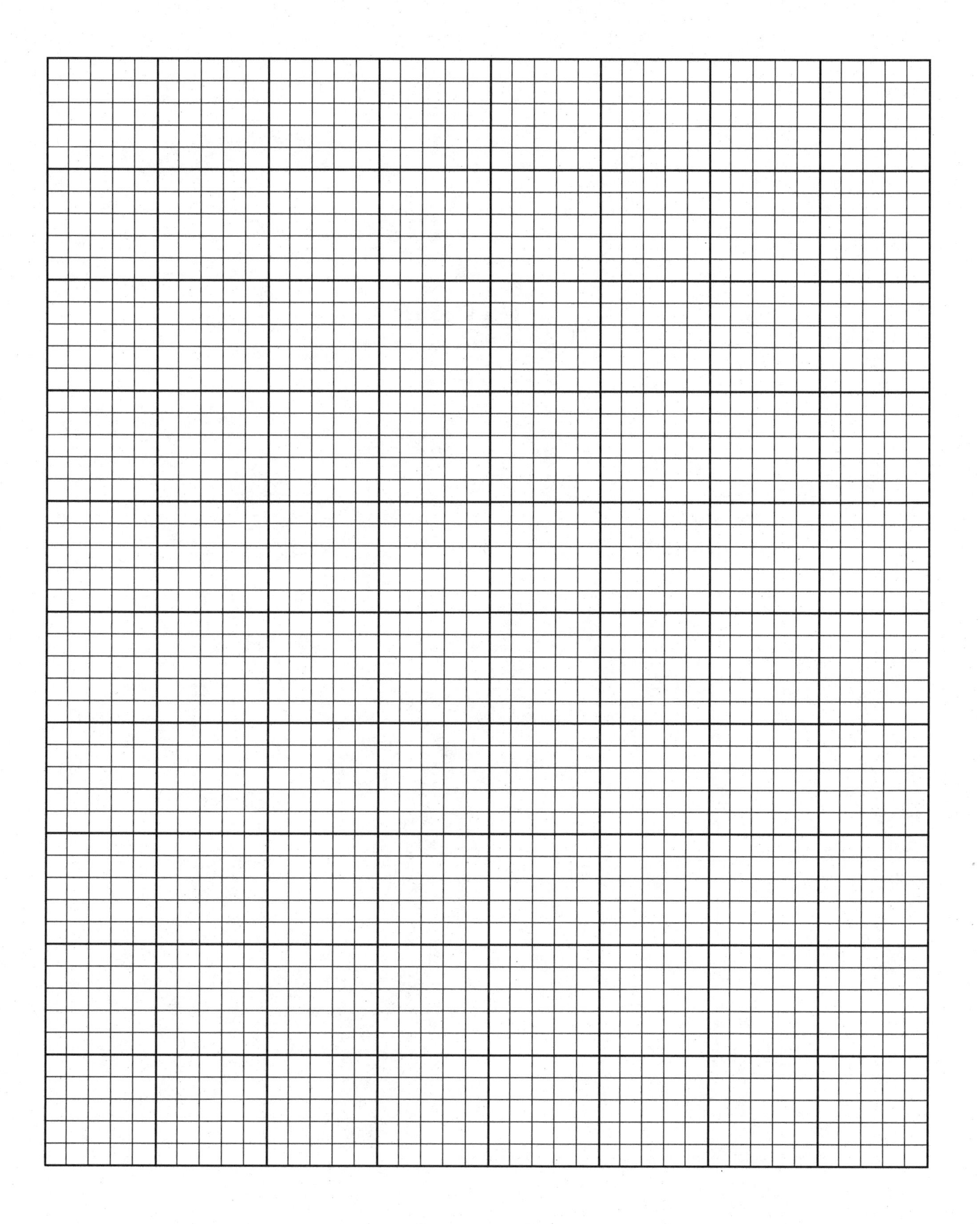

Circular Protractors

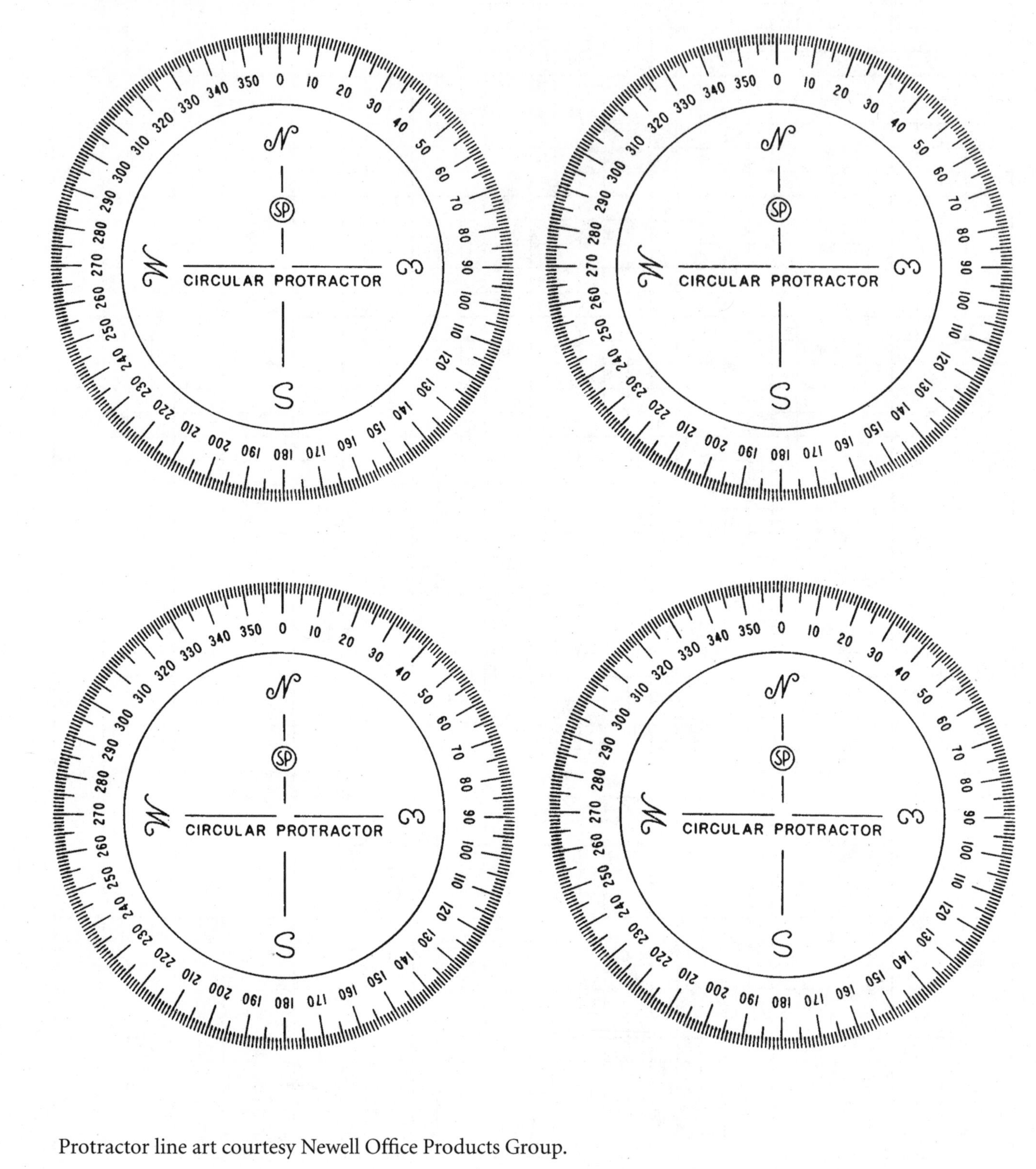

Protractor line art courtesy Newell Office Products Group.